APPLICATION
OF
INTEGRAL CALCULUS

DPH MATHEMATICS SERIES

APPLICATION OF INTEGRAL CALCULUS

By

A.K. Sharma

DISCOVERY PUBLISHING HOUSE
NEW DELHI-110002

Published by:

DISCOVERY PUBLISHING HOUSE
4383/4B, Ansari Road, Darya Ganj
New Delhi-110 002 (India)
Phone : +91-11-23279245; 23253475; 43596065
E-mail : discoverybooksindia@gmail.com
discoverypublishinghouse@gmail.com
namitwasan9@gmail.com
web : www.discoverypublishinggroup.com

First Published: **2005**

Reprinted: **2023**

ISBN: 978-81-7141-967-8

Application of Integral Calculus

Printed at:
Infinity Imaging Systems
Delhi

Preface

The book is written to meet the requirements of B.A., B.Sc., students of various Indian Universities. The subject matter is exhaustive and attempts are made to present things in an easy to understand style. In solving the questions, care has been taken to explain each step so that student can follow the subject matter themselves without even consulting others. A large numbers of solved and self practice problems (with hint and answer) have been included in each chapter to make students familiar with the types of questions set in various examinations.

I believe that the book will serve specific need of the students appearing in the examination of B.A./B.Sc., and other competitive examinations.

I gratefully acknowledge this inspiration, encouragement and valuable suggestions received from the well-wisher during the preparation of this book.

Suggestions and comments for the important of the book from reader will be thankfully received and duly incorporated in the subsequent editions.

A.K. Sharma

CONTENTS

1

Area of Curves
(Quadrature)

1.1 Definition

*The process of finding the area of any bounded portion of a curve is called **quadrature**.*

1.2 Areas of Curves Given by Cartesian Equations

If f(x) is a continuous and single valued function of x, then the area bounded by the curve $y = f(x)$, the axis of x and the ordinates $x = a$ and $x = b$ is

$$\int_a^b y\,dx, \text{ or } \int_a^b f(x)\,dx.$$

Proof:

Let CD be the arc of the curve $y = f(x)$ and AC and BD be two ordinates $x = a$ and $x = b$.

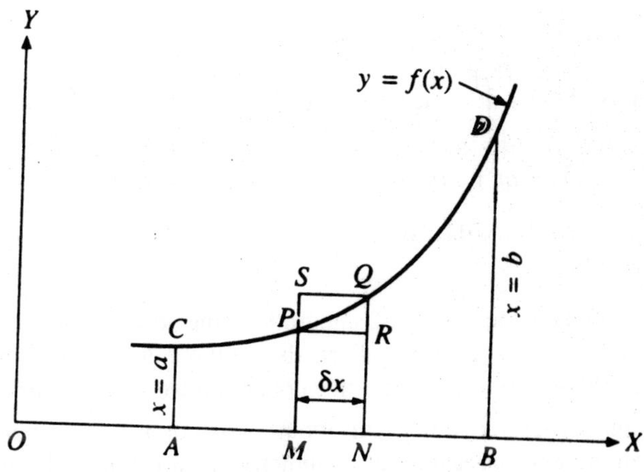

Fig. 1.1

Consider P(x, y) and Q(x + δx, y + δy), the two neighbouring points on the curve. Draw PM and QN perpendiculars to the axis of x, then

$$PM = y,\ QN = y + \delta y \text{ and } MN = dx.$$

Draw PR and QS perpendiculars to NQ and MP produced respectively. The area AMPC depends upon the position of P on the curve. Let A denote the area AMPC and A + δA be the area ANQC. Then the area MNQP = area ANQC – area AMPC

$$= A + \delta A - A = \delta A.$$

But clearly this area δA (*i.e.*, the area MNQP) lies in magnitude between the areas of the rectangles MNRP and MNQS.

Thus, we have

Area of the rectangle MNQS > δA > area of the rectangle MNRP

i.e., $$(y + \delta y)\delta x > \delta A > y\delta x$$

or $$y + \delta y > \frac{\delta A}{\delta x} > y.$$

Now as Q → P, δx → 0 and δy → 0. Therefore we have

$$\frac{dA}{dx} = y = f(x), \text{ or } dA = y\,dx.$$

Integrating both sides between the limits x = a and x = b, we have

$$\int_{x=a}^{x=b} dA = \int_a^b y\,dx$$

or $$[A]_{x=a}^{x=b} = \int_a^b y\,dx$$

or $$(\text{Area A when } x = b) - (\text{Area A when } x = a) = \int_a^b y\,dx$$

or $$\text{Area ABDC} - 0 = \int_a^b y\,dx$$

or $$\textbf{Area ABDC} = \int_a^b \mathbf{y\,dx} = \int_a^b \mathbf{f(x)\,dx.}$$

Similarly, it can be shown that the area bounded by the curve x = f (y), the axis of y and the abscissae y = a and y = b is

$$\int_a^b \mathbf{x\,dy,\ or} \int_a^b \mathbf{f(y)\,dy.}$$

Notes:

1. In choosing the limits of integration, the lower limit of integration should be taken as the smaller value of the independent variable while the greater value gives us the upper limit of integration.
2. If the curve is symmetrical about x-axis or y-axis or both, then we shall find the area of one symmetrical part and multiply it by the number of symmetrical parts to get the whole area.

1. *Rules for Tracing Cartesian Curves (Summary)*

(a) *Symmetry* : (i) If all the powers of y which occur in the equation of the curve are even, the curve is symmetrical about the axis of x. (ii) If all the powers of x are even the curve is symmetrical about the axis of y. (iii) If he powers of x and y are all even, the curve is symmetrical about both the axes. (iv) If on interchanging x and y the equation of the curve remains unchanged, then the curve is symmetrical about the straight line $y = x$. (v) If on putting $-x$ for x and $-y$ for y the equation of the curve is unaltered, then there is symmetry in opposite quadrants.

(b) *Asymptotes Parallel to the Axes* : The asymptotes parallel to the x-axis can be obtained by equating to zero the coefficient of the highest power of x in the equation of the curve. Similarly, the asymptotes parallel to the y-axis can be obtained by equating to zero the coefficient of the highest power of y in the equation of the curve.

(c) *Points of Inflexion* : While tracing the curve, if it appears that the curve possesses some points of inflexion, then they can be clearly located by putting (a^2y/dx^2) or (d^2x/dy^2) equal to zero and solving the resulting equation.

(d) Solve the equation of the curve for y or x whichever is convenient. Suppose we solve for y. Then put the equation of the curve in the form $y = f(x)$. Now we give values to x from 0 to ∞ and then from 0 to $-\infty$ and note the values of y.

(e) *Location at the Orign* : Observe whether the curve passes through the orign or not. If the point (0, 0) satisfies the equation of the curve, it passes through the origin. If the curve passes through the origin, then the tangents at the origin are obtained by equating to zero the lowest degree terms occurring in the equation of the curve.

(f) *Some Points on the Curve* : Find the point where the curve cuts the co-ordinate axes. In the equation of the curve, put $y = 0$ and solve for x. Again put $x = 0$ and solve for y. Thus, we get the points where the curve cuts the coordinate axes. Also find out the tangents at these points. Giving other suitable values to x find out some other points on the curve.

2. *Tracing of Polar Curve*

(a) Plot certain points on the curve. For this, give certain values to θ and find the corresponding values of r.

For this we must remember the following table:

θ =	0°	30°	60°	90°	120°	150°	180°	210°	240°	270°
$\sin\theta$ =	0	1/2	√3/2	1	√3/2	1/2	0	–1/2	–√3/2	–1
$\cos\theta$ =	1	√3/2	1/2	0	–1/2	–√3/2	–1	–√3/2	–1/2	0

(b) If the curve possesses an infinite branch we find its asymptotes.

(c) Find tan f = r (dq/dr). This gives the direction of the tangent at any point on the curve.

(d) *Symmetry. (i)* If the equation of the curve does not change by putting $-\theta$ for θ, then the curve is symmetrical about the initial line. (ii) If the equation of the curve remains unchanged by putting –r for r, then the curve is symmetrical about the pole and the pole is the centre of the curve. (iii) If the equation of the of the curve remains unchanged by changing θ into $(\pi - \theta)$, then curve is symmetrical about the line $\theta = \pi/2$ (*i.e.*, the y-axis). (iv) If the equation of the curve remains unchanged if q is changed into $1/2\pi - \theta$, then the curve is symmetrical about the line $\theta = \pi/4$.

(e) The curve will pall through the pole if for some value of θ the value of r comes out to be zero. Also if r = 0 when θ = a, then usually the line θ = a will be a tangent to the curve at the pole.

We should find the values of θ for which r = 0, or r is maximum, or r is minimum, or $r \to \infty$.

If r vanishes for more than one values of θ, the curve is said to have a loop. The limits for a loop are obtained by finding two consecutive values of θ for which r is zero.

1.3 Areas of Curves Given by Polar Equations

If $r = f(\theta)$ be the equation of a curve in polar coordinates where $f(\theta)$ is a single valued continuous function of θ then the area of the sector enclosed by the curve and the two radii vectors $\theta = \theta_1$ and $\theta = \theta_2$ $(\theta_1 < \theta_2)$, is equal to

$$\frac{1}{2}\int_{\theta=\theta_1}^{\theta_2} r^2\, d\theta.$$

Proof:

Let OAB be the area of the curve $r = f(\theta)$ between the radii vectors $\theta = \theta_1$ and $\theta = \theta_2$.

Let P(r, θ) be any point on the curve between A and B. Take a point Q($r + \delta r$, $\theta + \delta\theta$) on the curve very near to P and draw the radius vector

OQ. Let the sectorial areas AOP and AOQ be denoted by A and A + δA respectively.

Then the curvilinear area OPQO = A + δA – A = δA.

Also we have OP = r; OQ = r + δr and $\angle$ POQ = $\delta\theta$.

The area of the circular sector POQ'

$$= \frac{1}{2}(\text{radius} \times \text{arc}) = \frac{1}{2}r.$$

$$r\delta\theta = \frac{1}{2}r^2\,\delta\theta,$$

and the area of the circular sector P' OQ

$$= \frac{1}{2}(r + \delta r).\ (r + \delta r)$$

$$\delta\theta = \frac{1}{2}(r + \delta r)^2\,\delta\theta.$$

Now, area POQ' < area OPQ < area P' OQ,

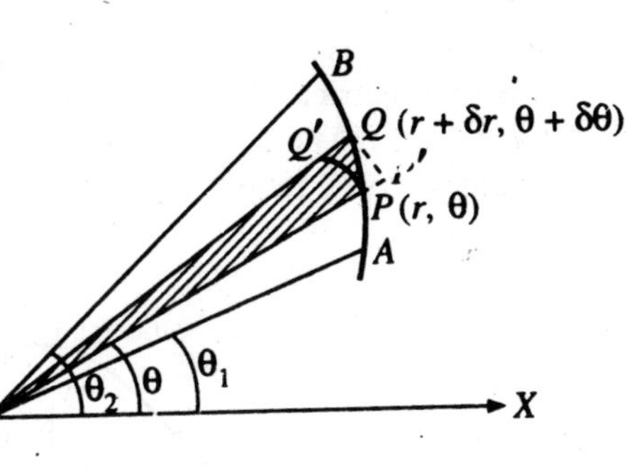

Fig 1.1(a)

i.e., $$\frac{1}{2}r^2\,\delta\theta < \delta A < \frac{1}{2}(r + \delta r)^2\,\delta\theta,$$

i.e., $$\frac{1}{2}r^2 < \delta A/\delta\theta < \frac{1}{2}(r + \delta r)^2.$$

Proceeding to limits as $\delta\theta \to 0$, we get

$$\frac{dA}{d\theta} = \frac{1}{2}r^2 \quad \text{or } dA = \frac{1}{2}r^2\,d\theta.$$

$$\therefore \qquad \left[A\right]_{\theta_1}^{\theta_2} = \int_{\theta_1}^{\theta_2} \frac{1}{2}r^2\,d\theta.$$

Now the L.H.S. = the value of A for θ equal to θ_2 – the value of A for θ equal to θ_1 = (the area AOB) – 0 = area AOB.

Hence the required area AOB

$$= \frac{1}{2}\int_{\theta_1}^{\theta_2} r^2\,d\theta.$$

Note: In some cases it is more convenient to find the required area by using double integration. In that case the area is given by

$$\int_{\theta=\theta_1}^{\theta_2}\int_{r=0}^{f(\theta)} r\,d\theta\,dr,\ (\theta_1 < \theta_2).$$

Remember : The number of loops in r = a cos nθ or r = a sin nθ is n or 2n according as n is odd or even.

MISCELLANEOUS EXAMPLES

Example 1:

Find the whole area of the curve $a^2y^2 = x^3 (2a - x)$.

Solution:

The given curve is $a^2y^2 = x^3 (2a - x)$.

It is symmetrical about x-axis and it cuts the x-axis at the points (0, 0) and 2a, 0). The curve does not exist for $x > 2a$ and $x < 0$. Thus the curve consists of a loop lying between $x = 0$ and $x = 2a$.

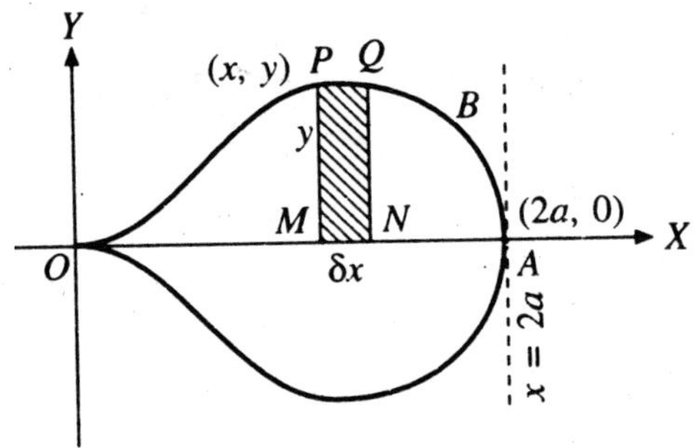

Fig. 1.2

$\therefore$ the required are

$$= 2 \times \text{area OBA}$$

$$= 2 \int_0^{2a} y\, dx$$

$$= 2 \int_0^{2a} \frac{x^{3/2} \sqrt{(2a - x)}}{a} dx, \text{ from (1).}$$

Now put $x = 2a \sin^2 \theta$

so that $dx = 4a \sin \theta \cos \theta \, d\theta$.

When $x = 0, \theta = 0$

and when $x = 2a, \theta = \frac{1}{2}\pi$.

$\therefore$ the required area

$$= \frac{2}{a} \int_0^{\pi/2} (2a)^{3/2} \sin^3 \theta . \sqrt{(2a)}.\cos\theta . 4a \sin\theta \cos d\theta$$

$$= 32a^2 \int_0^{\pi/2} \sin^4 \theta \cos^2 \theta \, d\theta$$

$$= 32a^2 . \frac{3.1.1}{6.4.2} . \frac{\pi}{2},$$ (by Walli's formula)

Example 2:

Find the whole area of the curve $a^2x^2 = y^3 (2a - y)$.

Solution:

The given curve is $a^2x^2 = y^3 (2a - y)$. ...(1)

It is symmetrical about y-axis and it cuts the y-axis at the points (0, 0) and (0, 2a). The curve does not exist for $y > 2a$ and $y < 0$.

$\therefore$ the required area = 2 × area OBA

$$= 2 \int_0^{2a} x\, dy$$

$$= 2 \int_0^{2a} \frac{y^{3/2} \sqrt{(2a - y)}}{a} dy,$$ from (1).

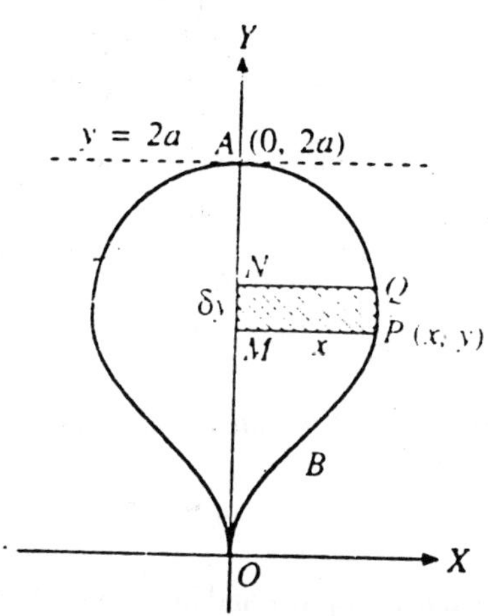

Fig. 1.3

Putting $y = 2a \sin^2 \theta$ we get the required area = pa^2. (Note that this is also the area of a circle of radius a).

Example 3:

Find the area bounded by the curve $xy^2 = 4a^2 (2a - x)$ and its asymptote.

Solution:

The given curve is symmetrical about the x-axis and cuts the x-axis at the point (2a, 0).

Equating to zero the coefficient of highest power of y in the equation of the curve we get x = 0 *i.e.*, the y-axis as the asymptote of the curve parallel to y-axis.

Hence the required area

$$= 2 \int_0^{2a} y \, dx$$

$$= 2\int_0^{2a} \frac{2a\sqrt{(2a-x)}}{\sqrt{x}} dx,$$

[∴ from the given equation of the curve, $y^2 = 4a^2 (2a - x)/x$].

Putting $x = 2a \sin^2\theta$

so that $dx = 4a \sin\theta \cos\theta \, d\theta$, we get the required area

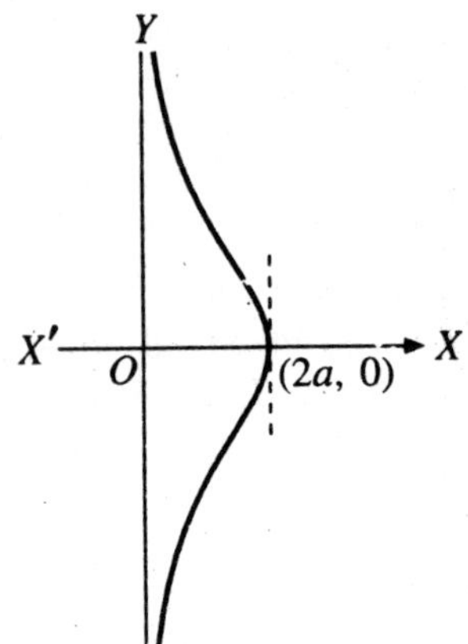

Fig 1.4

$$= 4a \int_0^{\pi/2} \frac{\sqrt{(2a)}\cos\theta . 4a \sin\theta \cos\theta d\theta}{\sqrt{(2a)}.\sin\theta}$$

$$= 16a^2 \int_0^{\pi/2} \cos^2\theta \, d\theta$$

$$= 16a^2 . \frac{1}{2}.\frac{1}{2}\pi, \text{ by Wallil's formula}$$

$$= 4\pi a^2.$$

Example 4:

Find the area enclosed by the curve $xy^2 = a^2 (a - x)$ and y-axis

Solution:

Do your self.

Here also y-axis is the asymptote and in place of 2a we have a *i.e.*, replace a/2 in place of a.

The required area = $4\pi. (a/2)^2 = \pi a^2$.

Example 5:

Trace the curve $a^2y^2 = a^2x^2 - x^4$ and find the whole area within it.

Solution :

The given curve is $a^2y^2 = x^2 (a^2 - x^2)$. Since in the equation of the curve, the powers of x and y are all even, therefore the curve is symmetrical about both the axes.

The curve passes through the origin and the tangents at the origin are $a^2y^2 - a^2x^2 = 0$ *i.e.*, $a^2(y^2 - x^2)$ 0 *i.e.*, $y^2 - x^2 = 0$ *i.e.*, $y = \pm x$.

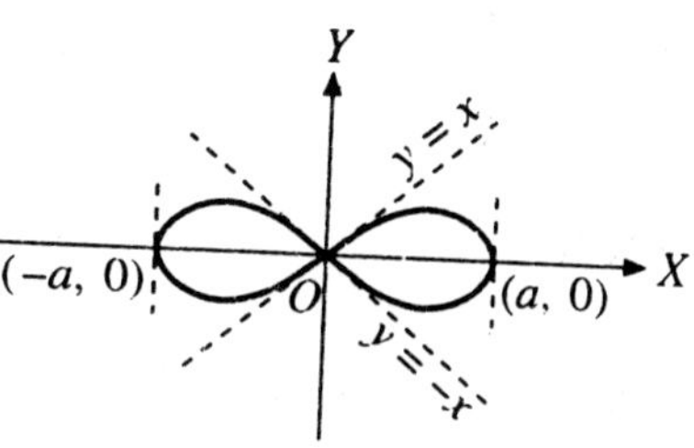

Fig 1.5

The curve cuts the axis of x where $y = 0$ *i.e.*, where $x^2(a^2 - x^2) = 0$ or $x = 0, \pm a$. Therefore the curve cuts the x-axis at (0, 0), (a, 0) and (–a, 0).

The curve intersects the y-axis only at the origin.

Tangent at (a, 0)

Shifting the origin to the point (a, 0) the equation of the curve becomes

$$a^2y^2 = (x + a)^2 \{a^2 - (x + a)^2\} = (x + a)^2 (-x^2 - 2ax).$$

Equating to zero the lowest degree terms, we get $x = 0$ as the tangent at the new origin. Thus new y-axis is tangent at the new origin.

Solving the equation of the curve for y, we get

$$y^2 = \frac{x^2(a^2 - x^2)}{a^2}.$$

When $x = 0,\ y^2 = 0.$

When $x = a,\ y^2 = 0.$

When $0 < x < a, \quad y^2$ is +ive.

Therefore the curve exists in the region $0 < x < a$.

When $x > a$, y^2 is –ive. Therefore the curve does not exist in the region $x > a$.

Hence the curve is as shown in the figure and it consists of two equal loops.

By symmetry, the whole area within the curve

$$= 4 \times \text{area of half a loop}$$

$$= 4 \int_0^a y\,dx = 4 \int_0^a \frac{x\sqrt{(a^2 - x^2)}}{a} dx,$$

putting for y from the given equation of the curve

$$= 4 \int_0^{\pi/2} \sin\theta.\ a\cos\theta.\ a\cos\theta\ d\theta.$$

putting $x = a\sin\theta$

so that $dx = a\cos\theta\ d\theta$

$$= 4a^2 \int_0^{\pi/2} \cos^2\theta \sin\theta \, d\theta$$

$$= 4a^2 \cdot \frac{1}{3.1} .1. \qquad \text{(by Walli's formula} = \frac{4}{3}a^2)$$

Example 6:

Find the area of the curve $y^2 = x^2 (4 - x^2)$.

Solution:

Do your self. The required area

$$= 4 \int_0^2 x \sqrt{(4 - x^2)} \, dx = \frac{32}{3}.$$

Example 7:

Prove that the area of a loop of he curve $a^4y^2 = x^4(a^2 - x^2)$ is $\pi a^2/8$.

Solution:

The curve is symmetrical about both the axes. Putting y = 0 in the given equation of the curve, we get $x^4 (a^2 - x^2) = 0$ *i.e.*, $x = 0$, $x = \pm a$. Thus, the above curve will have a loop between $x = 0$ and $x = a$. By symmetry, the area of a loop

$$= 2 \int_0^a y \, dx = 2 \int_0^a \frac{x^2 \sqrt{(a^2 - x^2)}}{a^2} dx,$$

Putting for y from the given equation of the curve

$$= \frac{2}{a^2} \int_0^{\pi/2} a^2 \sin^2\theta \sqrt{(a^2 - a^2 \sin^2\theta)}. \, a \cos\theta \, d\theta$$

putting $x = a \sin\theta$ so that $dx = a \cos\theta \, d\theta$

$$= 2a^2 \int_0^{\pi/2} \sin^2\theta \cos^2\theta \, d\theta$$

$$= 2a^2. \frac{1.1}{4.2} \cdot \frac{1}{2}\pi, \text{ by Walli's formula}$$

$$= \pi a^2/8.$$

Example 8:

Show that the whole area of the curve $a^4y^2 = x^5 (2a - x)$ is to that of the circle whose radius is a, as 5 to 4.

Solution:

The given curve is symmetrical about x-axis. It passes through the origin and the tangents at the origin are $a^4y^2 = 0$ *i.e.*, $y^2 = 0$ *i.e.*, $y = 0$, $y = 0$.

The curve cuts the x-axis at the points (0, 0) and (2a, 0). It intersects the y-axis only at the origin. When $0 < x < 2a$, y^2 is +ive so that the curve exists in this region. When $x > 2a$, y^2 is –ive so that the curve does not exist in this region. When $x < 0$, y^2 is –ive so that curve does not exist in this region.

Thus, the given curve consists of a loop lying between $x = 0$ and $x = 2a$. Hence, ıhe whole area of this curve

$$= 2\int_0^{2a} y\,dx = 2\int_0^{2a} \frac{x^{5/2}(2a-x)^{1/2}}{a^2}dx,$$

Putting for y from the given equation of the curve

$$= 2\int_0^{\pi/2} \frac{(2a)^{5/2}\sin^5\theta.(2a)^{1/2}\cos.\theta 4a\sin\theta\cos\theta d\theta}{a^2},$$

Putting $x = 2a\sin^2\theta$

so that $dx = 2a.2\sin\theta\cos\theta\,d\theta$

$$= 64a^2\int_0^{\pi/2}\sin^6\theta\cos 2\theta\,d\theta$$

$$= 64a^2\frac{\Gamma\left(\frac{7}{2}\right)\Gamma\left(\frac{3}{2}\right)}{2\Gamma(5)} = 64a^2\frac{\frac{5}{2}.\frac{3}{2}.\frac{1}{2}.\sqrt{\pi}.\frac{1}{2}\sqrt{\pi}}{2.4.3.2.1} = \frac{5\pi a^2}{4}.$$

Also the area of the circle of ratius is πa^2.

$$\therefore \quad \frac{\text{Area of the curve}}{\text{Area of the circle}} = \frac{5\pi a^2/4}{\pi a^2} = \frac{5}{4}.$$

Example 9:

Find the area between the curve $y^2(a-x) = x^3$ **(cissoid)** *and its asymptotes. Also find the ratio in which the ordinate $x = a/2$ divides the area.*

Solution:

Equating to zero the coefficient of the highest power of y, we get $a - x = 0$ *i.e.*, $x = a$ as an asymptote of the curve parallel to the y-axis.

Let A be the whole area between the curve and its asymptote. Then

$$A = 2 \times (\text{area in the first quadrant})$$

$$= 2\int_0^a y\,dx = 2\int_0^a \frac{x^{3/2}}{\sqrt{(a-x)}}dx,$$

putting for y from the given equation of the curve

$$= 2\int_0^{\pi/2} \frac{a^{3/2}\sin^3\theta . 2a\sin\theta\cos\theta}{\sqrt{(a)}.\cos\theta} d\theta,$$

putting $x = a\sin^2\theta$ so that $dx = a.2\sin\theta\cos\theta\, d\theta$

$$= 4a^2\int_0^{\pi/2}\sin^4\theta\, d\theta = 4a^2 . \frac{3.1}{4.2}.\frac{\pi}{2} = \frac{3\pi a^2}{4}. \quad ...(1)$$

Now let A_1 be the area of the portion between $x = 0$ and $x = \frac{1}{2}a$. Then

$$A_1 = 2\int_0^{a/2} y\, dx = 2\int_0^{a/2}\frac{x^{3/2}}{\sqrt{(a-x)}}dx,$$

putting for y from the given equation of the curve

$$= 4a^2\int_0^{\pi/4}\sin^4\theta\, d\theta, \text{ putting } x = a\sin^2\theta, \text{ etc.}$$

$$= a^2\int_0^{\pi/4}(2\sin^2\theta)^2\, d\theta = a^2\int_0^{\pi/4}(1-\cos 2\theta)^2\, d\theta$$

$$= \frac{1}{2}a^2\int_0^{\pi/2}(1-\cos\phi)^2\, d\phi,$$

[putting $2\theta = \phi$ so that $2\, d\theta = d\phi$ and adjusting the limits]

$$= \frac{1}{2}a^2\int_0^{\pi/2}(1 - 2\cos\phi + \cos^2\phi)\, d\phi$$

$$= \frac{1}{2}a^2\left[\int_0^{\pi/2}d\phi - 2\int_0^{\pi/2}\cos\phi d\phi + \int_0^{\pi/2}\cos^2\phi d\phi\right]$$

$$= \frac{1}{2}a^2\left[\{\phi\}_0^{\pi/2} - 2\{\sin\phi\}_0^{\pi/2} + \frac{1}{2}.\frac{\pi}{2}\right]$$

$$= \frac{1}{2}a^2\left[\frac{1}{2}\pi - 2.1 + \frac{1}{4}\pi\right]$$

$$= \frac{1}{2}a^2\left[\frac{3\pi}{4} - 2\right] = \frac{1}{8}a^2(3\pi - 8).$$

Now let A_2 be the area of the portion of the curve lying between $x = \frac{1}{2}a$ and $x = a$.

The $A_2 = A - A_1 = \frac{3}{4}\pi a^2 - \frac{1}{8}a^2(3\pi - 8) = \frac{1}{8}a^2(3\pi + 8)$.

$\therefore$ required ratio $= \dfrac{A_1}{A_2} = \dfrac{(a^2/8)(3\pi - 8)}{(a^2/8)(3\pi + 8)} = \dfrac{3\pi - 8}{3\pi + 8}$.

Example 10:

Find the whole area between the curve $x^2y^2 = a^2 (y^2 - x^2)$ and its asymptotes.

Solution:

The given curve is symmetrical about both the axes and passes through the origin. The tangents at (, 0) are given by $y^2 - x^2 = 0$ *i.e.*, $y = \pm x$ are tangents at the origin.

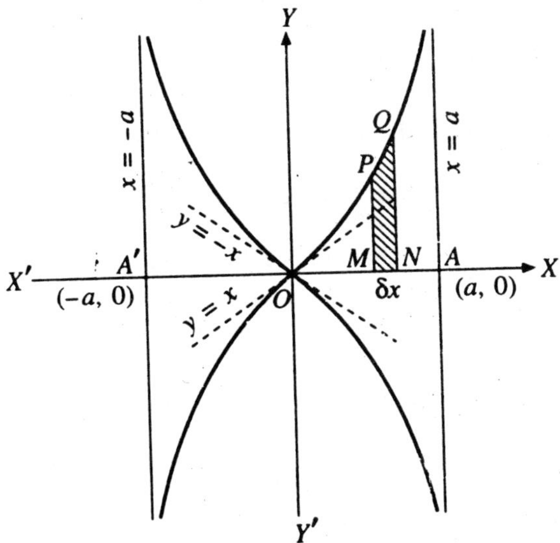

Fig. 1.6

Equating to zero the coefficient of the highest power of y (*i.e.*, of y^2) the asymptotes parallel to y-axis are given by $x^2 - a^2 = 0$ *i.e.*, $x = \pm a$.

The asymptotes parallel to x-axis are given by $y^2 + a^2 = 0$ which gives two imaginary asymptotes.

∴ the required area = 4 × area lying in the first quadrant

$$= 4 \int_0^a y\,dx = 4 \int_0^a \sqrt{\left(\frac{a^2x^2}{a^2 - x^2}\right)}\,dx,$$

[∴ from the equation of the given curve, $y^2 = a^2x^2/(a^2 - x^2)$]

$$= 4 \int_0^a \frac{axdx}{\sqrt{(a^2 - x^2)}} = -2a \int_0^a \frac{-2xdx}{\sqrt{(a^2 - x^2)}}$$

$$= -2a\left[\frac{(a^2-x^2)^{1/2}}{1/2}\right]_0^a$$

$$= -4a\,[0-a] = 4a^2.$$

Example 11:

Find the area of the loop of the curve

$$y^2(a-x) = x^2(a+x).$$

Solution:

The given curve is symmetrical about the x-axis and cut the x-axis at the points (0, 0) and (–a, 0).

The tangents at (0, 0) are $y^2 = x^2$ *i.e.*, $y = \pm x$.

Clearly there is a loop which lies between $x = -a$ and $x = 0$.

∴ the required are of the loop = 2 × area of the upper half of the loop

$$= 2\int_{-a}^{0} y\,dx = 2\int_{-a}^{0} -x.\frac{\sqrt{(a+x)}}{\sqrt{(a-x)}}dx,$$

putting for y for the upper half of the loop from the given equation of the curve

$$= 2\int_{-a}^{0} -x\frac{(a+x)}{\sqrt{(a^2-x^2)}}dx,$$

multiplying the numerator and the denominator by $\sqrt{(a+x)}$.

Now put $x = -a\sin\theta$ so that $dx = -a\cos\theta\,d\theta$.

When $x = -a$, $\theta = \pi/2$ and when $x = 0$, $\theta = 0$.

∴ the required area

$$= 2\int_{\pi/2}^{0} -(-a\sin\theta)\frac{(a-a\sin\theta)}{\sqrt{(a^2-a^2\sin^2\theta)}}.(-a\cos\theta)d\theta$$

$$= -2a^2\int_{\pi/2}^{0} \sin\theta\,(1-\sin\theta)\,d\theta$$

$$= 2a^2\int_{0}^{\pi/2}(\sin\theta-\sin^2\theta)\,d\theta$$

$$= 2a^2\left[1-\frac{1}{2}.\frac{1}{2}\pi\right], \qquad \text{(by Walli's formula)}$$

$$= 2a^2\left(1-\frac{1}{4}\pi\right) = \frac{1}{2}a^2(4-\pi).$$

Example 12:

Trace the curve $4y^2(a+x) = (a-x)^3$ *and find the area between the curve and its asymptotes.*

Solution:

The curve is symmetrical about x-axis. It does not pass through the origin. Putting x = 0 in the equation of the curve, we get $y = \pm a/2$ and putting y = 0 in it we get x = a. Thus, the curve cuts the y-axis at the points $(0, \pm 1/2a)$ and it cuts the x-axis at the point (a, 0).

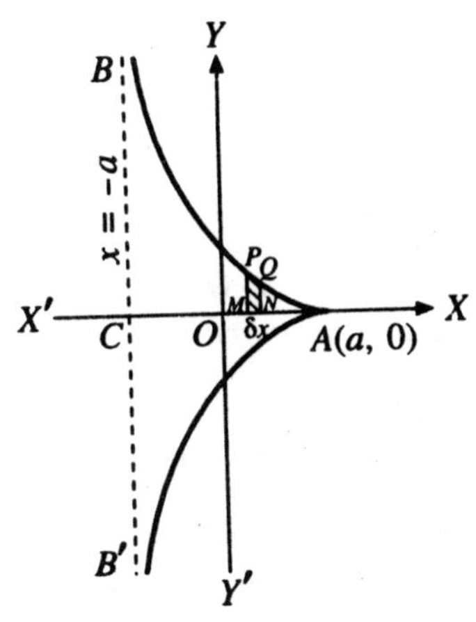

Fig 1.7

Equating to zero the coefficient of the highest power of y the asymptote parallel to y-axis is a + x = 0 *i.e.*, x = −a.

The equation of the curve can be written as $y^2 = (a - x)^3/\{4(a + x)\}$ which shows that for x > a, y is imaginary *i.e.*, the curve does not exist for x > a.

Thus, the shape of the curve is as shown in the fig. Now the required area = 2 × area lying above the x-axis.

$$= 2\int_{-a}^{a} y\,dx = 2\cdot\int_{-a}^{a} \frac{1}{2}\frac{(a-x)^{3/2}}{\sqrt{(a+x)}}\,dx,$$

puting for y from the equation of the curve

$$= \int_{-a}^{a} \frac{(a-x)^2}{\sqrt{(a^2+x^2)}}\,dx,$$

multiplying the Nr. and the Dr. by $\sqrt{(a - x)}$

$$= \int_{-\pi/2}^{\pi/2} \frac{(a - a\sin\theta)^2}{\sqrt{(a^2 + a^2\sin^2\theta)}}.a\cos\theta\,d\theta,$$

putting $x = a\sin\theta$ so that $dx = a\cos\theta\,d\theta$

$$= a^2\int_{-\pi/2}^{\pi/2} (1 - \sin\theta)^2\,d\theta$$

$$= a^2\int_{-\pi/2}^{\pi/2} (1 - \sin\theta + \sin^2\theta)\,d\theta$$

$$= a^2\int_{-\pi/2}^{\pi/2} \{1 - 2\sin\theta + 1/2\,(1 - \cos 2\theta)\}\,d\theta$$

$$= a^2 \left[\frac{3}{2}\theta + 2\cos\theta - \frac{1}{4}\sin 2\theta\right]_{-\pi/2}^{\pi/2}$$

$$= a^2\left(\frac{3\pi}{2}\right) = \frac{3}{2}\pi a^2.$$

Example 13:

Trace the curve $y^2 (a + x) = (a - x)^3$ and find the area between the curve and its asymptotes.

Solution:

Do your self. The required area = $3\pi a^2$.

Example 14

Find the area included between the curves $y^2 = 4ax$ and $x^2 = 4by$

Solution:

Solving the equations of the two given curves, we have $y^4 = 16a^2 (4by) = 64a^2by$.

$\therefore$ $y (y^3 - 64 a^2b) = 0$,

giving $y = 0, 4a^{2/3} b^{1/3}$.

When $y = 0$, $x = 0$ and when $y = 4a^{2/3} b^{1/3}$, $x = 4a^{1/3} b^{2/3}$

Hence, the points of intersection of the given curves are O (0, 0) and A $(4a^{1/3} b^{2/3}, 4b^{1/3} b^{2/3})$.

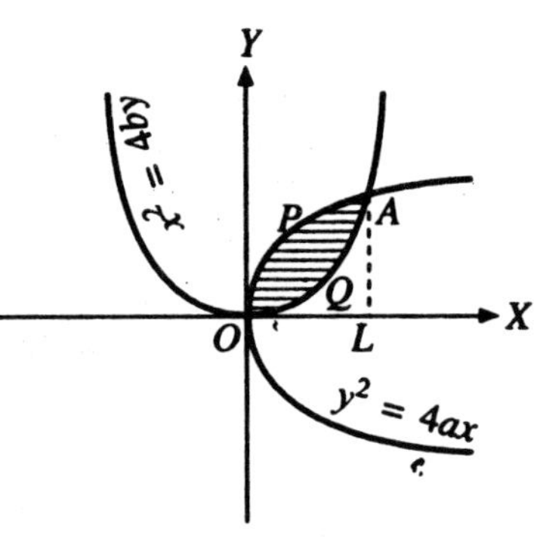

Fig 1.8

$\therefore$ the required area (*i.e.*, the shaded area)

$$= \text{area OPAL} - \text{area OQAL}$$

$$= \int_0^{4a^{1/3}b^{2/3}} y\,dx, \text{ form the curve } y^2 = 4ax$$

$$- \int_0^{4a^{1/3}b^{2/3}} y\,dx, \text{ form the curve } x^2 = 4by$$

(Note that for the required area x varies from 0 to $4a^{1/3} b^{2/3}$)

$$= \int_0^{4a^{1/3}b^{2/3}} \sqrt{(4ax)}\,dx - \int_0^{4a^{1/3}b^{2/3}} \left(\frac{x^2}{4b}\right)dx$$

$$- 2\sqrt{a}\left[\frac{2x^{3/2}}{3}\right]_0^{4a^{1/3}b^{2/3}} - \frac{1}{4b}\left[\frac{x^3}{3}\right]_0^{4a^{1/3}b^{2/3}}$$

$$= \frac{4\sqrt{a}}{3}[8\sqrt{(a)}.b] - \frac{1}{12b}(64\ ab^2)$$

$$= \frac{32}{3}ab - \frac{16}{3}ab = \frac{16}{3}ab.$$

Example 15:

Find the area of a loop of the curve r = a sin 3θ.

Solution:

The given curve is not symmetrical about the initial line. We have $r = 0$ when $\sin 3\theta = 0$ *i.e.*, $3\theta = 0, \pi$ *i.e.*, $\theta = 0, 1/3\pi$. Thus, two consecutive values of θ for which r is zero are and $1/3\pi$. Therefore, one loop of the curve lies between $\theta = 0$ and $1/3\pi$. In all there are three loops as shown in the figure. For the first loop θ varies from $\theta = 0$ to $\theta = 1/3\pi$.

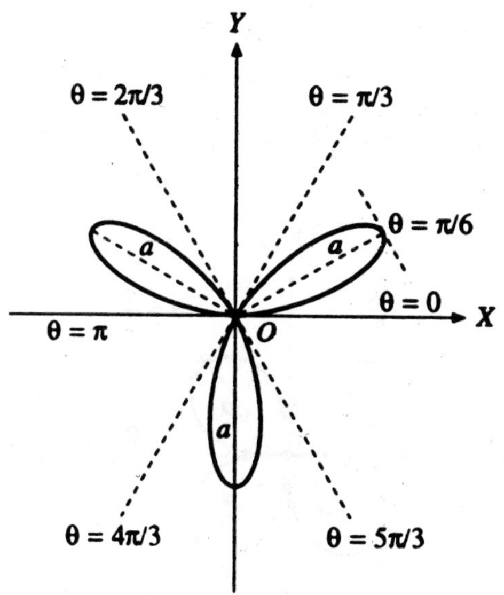

Fig 1.9

Hence the area of a loop

$$= \frac{1}{2}\int_0^{\pi/3} r^2\, d\theta$$

$$= \frac{1}{2}\int_0^{\pi/3} a^2 \sin^2 3\theta\, d\theta, \qquad [\because r = a \sin 3\theta]$$

$$= \frac{a^2}{4}\int_0^{\pi/3} (1 - \cos 6\theta)\, d\theta$$

$$= \frac{1}{4} a2 \left[\theta - \frac{1}{6}\sin 6\theta\right]_0^{\pi/3}$$

$$= \frac{a^2}{4}\left[\frac{\pi}{3}\right] = \frac{\pi a^2}{12}.$$

Note: Also whole area of the curve r = a sin 3q = 3 ×a area of one loop

$= 3 \times (1/12)\, \pi\, a^2 = 1/4\pi\, a^2$.

Remark:

The above curve is a particular case of the curves of the type r = a sin nθ which have n loop when n is odd and 2n loop when n is even.

Example 16:

Find the whole curve is r = a sin 2θ.

Solution:

Here the given curve is r = a sin 2θ.

Comparing with r = a sin nθ we observe that n = 2 (*i.e.*, even), therefore the curve has four loops. This curve is not symmetrical about the initial line. Putting r = 0 *i.e.*, 2θ = 0, r *i.e.*, θ = 0, 1/2π.

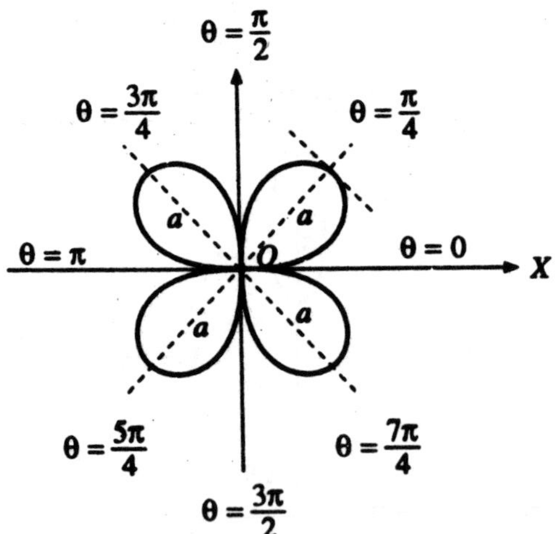

Fig 1.10

Thus two consecutive values of θ for which r is zero are 0 and $1/2\pi$

Therefore for one loop of the curve θ varies from 0 to $1/2\pi$.

Now the whole area of the curve = 4 × area of one loop

$$= 4\int_0^{\pi/2} \frac{1}{2} r^2 \, d\theta = 2\int_0^{\pi/2} a^2 \sin^2 2\theta \, d\theta$$

$$= a^2 \int_0^{\pi/2} 2 \sin^2 2\theta \, d\theta = a^2 \int_0^{\pi/2} (1 - \cos 4\theta) \, d\theta$$

$$= a^2 \left[\theta - \frac{\sin 4\theta}{4}\right]_0^{\pi/2} = \frac{1}{2}\pi \, a^2.$$

Example 17:

Find the whole area of the curve r = a cos 2θ.

Solution:

Comparing the given equation with r = a cos nθ we observe than n = 2 (*i.e.*, even), therefore the number of loops = 2n = 2 × 2 = 4.

This curve is symmetrical about the initial line.

Putting r = 0, we get cos 2θ = 0 or 2θ = ± $1/2\pi$ or θ = ± $1/4\pi$ *i.e.*, for the first loop θ varies from $-\pi/4$ to $\pi/4$ and this loop is symmetrical about the initial line θ = 0.

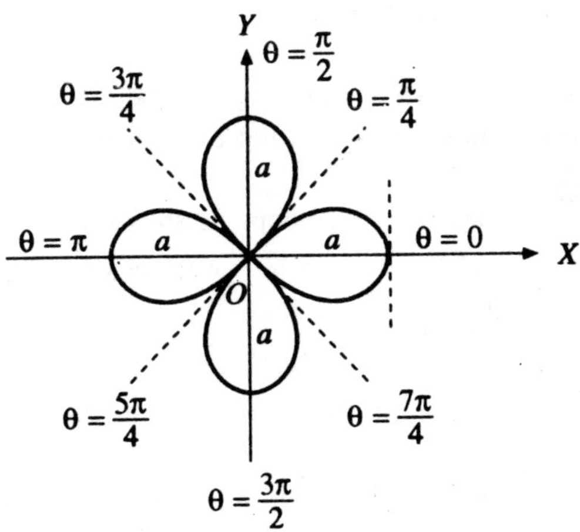

Fig 1.11

∴ whole area of the curve = 4 × area of one loop

$$= 4 \times 2\int_0^{\pi/2} \frac{1}{2} r^2 \, d\theta = 4\int_0^{\pi/4} a^2 \cos^2 2\theta \, d\theta.$$

Now put $2\theta = t$ so that $2\, d\theta = dt$.

When $\theta = 0$, $t = 0$ and

when $\theta = \frac{1}{4}\pi$, $t = \frac{1}{2}\pi$.

$$\therefore \text{ The required area} = 4a^2\int_0^{\pi/2} \cos^2 t \,.\frac{1}{2} dt$$

$$= 2a^2\int_0^{\pi/2} \cos^2 t \, dt$$

$$= 2a^2 \,.\frac{1}{2}.\frac{\pi}{2} = \frac{1}{2}\pi \, a^2.$$

Example 18:

Find the area of the curve $r^2 = a^2 \cos 2\theta$.

Solution:

The given curve is symmetrical about the initial line $\theta = 0$ and about the pole. Putting $r = 0$ in the give equation of the curve, we get $\cos 2\theta = 0$ or $2\theta = \pm\ 1/2\pi$ or $\theta = \pm\ 1/4\pi$. Thus, two consecutive values of θ for which r is zero are $-1/4\ \pi$ and $1/4\ \pi$. Therefore one loop of the curve θ varies from $-\pi/4$ to $\pi/4$.

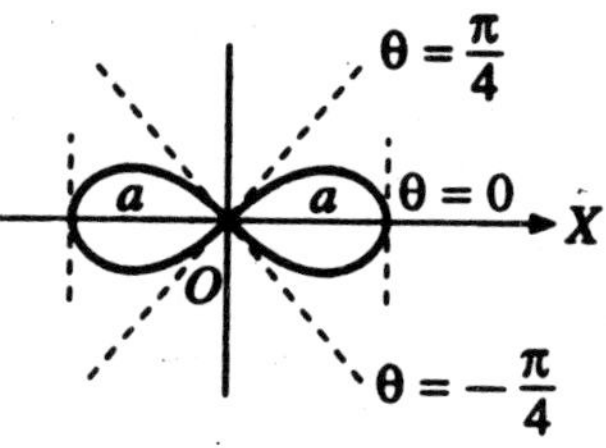

Fig 1.12

When $1/2\pi < 2\theta < 3/2\pi$ *i.e.*, $1/4\pi < \theta < 3/4\pi$, r^2 is negative *i.e.*, r is imaginary. There fore this curve does not exist in the region $1/4\pi < \theta < 3/4\pi$.

Hence this curve has only two loops as shown in the figure.

$\therefore$ whole area of the curve = 2 × area of one loop

$$= 2\int_{-\pi/4}^{\pi/4} \frac{1}{2} r^2 \, d\theta$$

$$= \int_{-\pi/4}^{\pi/4} a^2 \cos 2\theta \, d\theta, \ (\because r^2 = a^2 \cos 2\theta)$$

$$= 2a^2\int_0^{\pi/4} \cos 2\theta \, d\theta, \quad \text{(by a property of definite integrals)}$$

$$= 2a^2\left[\frac{\sin 2\theta}{2}\right]_0^{\pi/4}$$

$$= \frac{2a^2}{2} = a^2.$$

Example 19:

Find the common area between the curves $y^2 = 4ax$ and $x^2 = 4ay$.

Solution:

Here b = a. Solving the given equations, we get the points of intersection as (0, 0) and (4a, 4a) and hence required area $= \frac{16}{3}a^2$.

By double integration. The required area

$$= \int_{x=0}^{4a} \int_{y=(x^2/4a)}^{2\sqrt{(ax)}} dx\ dy$$

$$= \int_0^{4a} [y]_{(x^2/4a)}^{2\sqrt{(ax)}} dx$$

$$= \int_0^{4a} \left[2\sqrt{(ax)} - \frac{x^2}{4a}\right] dx$$

$$= 2\sqrt{(a)}\left[\frac{2}{3}x^{3/2}\right]_0^{4a} - \frac{1}{12a}\left[x^3\right]_0^{4a}$$

$$= \frac{4}{3}\sqrt{(a)}.\ (4a)^{3/2} - \left[\frac{(4a)^3}{12a}\right]$$

$$= \frac{32}{3}a^2 - \frac{16}{3}a^2 = \frac{16}{3}a^2.$$

Example 20:

The curves $y = 4x^2$ and $y^2 = 2x$, meet at the origin O and the point A, forming a loop. Show that the straight line OA divides the loop into two parts of equal area.

Solution:

Solving the equation of the two given curves, we have

$16x^4 = 2x$ or $16x^4 - 2x = 0$

i.e., $x = 0$ and $x = 1/2$.

When $x = 0$, $y = 0$

and when $x = 1/2$, $y = 1$.

Thus, the points of intersection are (0, 0) and $\left(\frac{1}{2}, 1\right)$.

The equation of the line OA is $y - 0 = \dfrac{1-0}{\frac{1}{2}-0}(x - 0)$ *i.e.*, $y = 2x$.

Now the area between the parabola $y^2 = 2x$ and the line $y = 2x$ = $\int_0^{1/2} y\,dx$, form the parabola $y^2 = 2x$ − $\int_0^{1/2} y\,dx$, from the line $y = 2x$

$$= \int_0^{1/2} \sqrt{(2x)}\,dx - \int_0^{1/2} 2x\,dx$$

$$= \left[\frac{2\sqrt{2}}{3}x^{3/2} - x^2\right]_0^{1/2}$$

$$= \frac{2\sqrt{2}}{3}\cdot\frac{1}{2\sqrt{2}} - \frac{1}{4} = \frac{1}{3} - \frac{1}{4} = \frac{1}{12}. \qquad ...(1)$$

Again the area between the parabola $y = 4x^2$ and the line $y = 2x$

$= \int_0^{1/2} y\,dx$, from the line $y = 2x$

$- \int_0^{1/2} y\,dx$, from the parabola $y = 4x^2$

$$= \int_0^{1/2} 2x\,dx - \int_0^{1/2} 4x^2\,dx = \left[x^2\right]_0^{1/2} - \frac{4}{3}\left[x^3\right]_0^{1/2}$$

$$= \frac{1}{4} - \frac{1}{6} = \frac{1}{12}. \qquad ...(2)$$

From (1) and (2) we observe that the straight line OA divides the loop into two parts of equal area.

Example 21:

Find the area included between $y^2 = 4ax$ and $y = mx$.

Solution:

Solving the equation of the parabola $y^2 = 4ax$ and the equation of the line $y = mx$ for x, we get $m^2x^2 = 4ax$ or $x(m^2x - 4a) = 0$.

This gives $x = 0$ or $x = 4a/m^2$.

Thus the two curves cot at the points where $x = 0$ and $x = 4a/m^2$.

∴ the required area

$= \int_0^{4a/m^2} y\,dx$, from the curve $y^2 = 4ax$

$- \int_0^{4a/m^2} y\,dx$, from the st. line $y = mx$

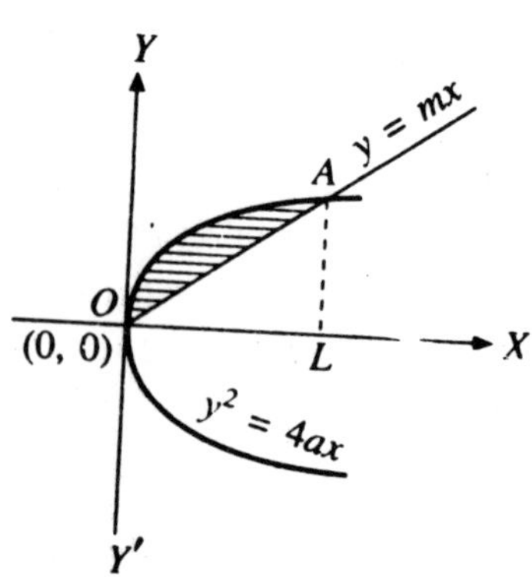

Fig 1.13

$$= \int_0^{4a/m^2} \sqrt{(4ax)}\, dx - \int_0^{4a/m^2} mx\, dx$$

$$= 2\sqrt{(a)}\left[\frac{2}{3}x^{3/2}\right]_0^{4a/m^2} - m\left[\frac{1}{2}x^2\right]_0^{4a/m^2}$$

$$= \frac{4\sqrt{a}}{3}\left(\frac{4a}{m^2}\right)^{3/2} - \frac{1}{2}m\left(\frac{4a}{m^2}\right)^2$$

$$= \frac{32a^2}{3m^3} - \frac{8a^2}{m^3} = \frac{8a^2}{3m^3}.$$

Example 22:

Find the area included between $y^2 = 9x$ and $y = x$.

Solution:

Do your self.

Here $m = 1$ and $a = 9/4$.

Example 23:

Find the area of the segment cut off from the parabola $y^2 = 2x$ by the straight line $y = 4x - 1$.

Solution:

The given curves are

$$y^2 = 2x, \quad ...(1)$$

and

$$y = 4x - 1 ...(2)$$

The two curves have been shown in the figure. Solving (1) and (2) for y we have

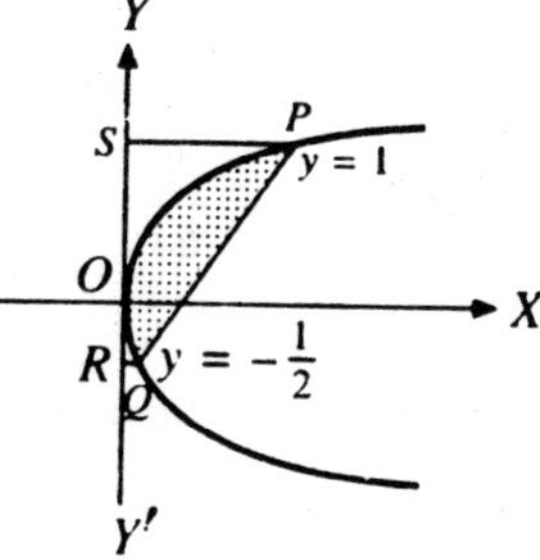

Fig 1.14

$$y^2 = 2.\frac{1}{4}(y + 1)$$

or $$2y^2 - y - 1 = 0$$

or $$(y - 1)(2y + 1) = 0.$$

$$\therefore\ y = -1/2,\ 1.$$

Thus, the curves (1) and (2) intersect at the points where $y = -1/2$ and $y = 1$.

Now the required area of the segment POQ (*i.e.*, the dotted area) = the area bounded by the st. line $y = 4x - 1$ and the y-axis from

$$y = -\frac{1}{2} \text{ to } y = 1$$

– the area bounded by the parabola $y^2 = 2x$ and the y-axis

$$\text{from } y = -\frac{1}{2} \text{ to } y = 1$$

$$= \int_{-1/2}^{1} x\, dy, \text{ from (2)} - \int_{-1/2}^{1} x\, dy, \text{ from (1)}$$

$$= \int_{-1/2}^{1} \frac{1}{4}(y + 1)\, dy - \int_{-1/2}^{1} \frac{1}{2} y^2\, dy$$

$$= \frac{1}{4}\left[\frac{1}{2}y^2 + y\right]_{-1/2}^{1} - \frac{1}{6}\left[y^3\right]_{-1/2}^{1}$$

$$= \frac{1}{4}\left[\frac{3}{2} - \left(\frac{1}{8} - \frac{1}{2}\right)\right] - \frac{1}{6}\left(1 + \frac{1}{8}\right)$$

$$= \frac{1}{4}\left(\frac{3}{2} + \frac{3}{8}\right) - \frac{1}{6}.\frac{9}{8} = \frac{1}{4}.\frac{15}{8} - \frac{1}{6}.\frac{9}{8}$$

$$= \frac{15}{32} - \frac{3}{16} = \frac{9}{32}.$$

Example 24:

Find the area of the segment cut off from the parabola $y^2 = 4x$ by the line $y = 8x - 1$.

Solution:

Do your self.

Here the points of intersection are $= \left(\frac{1}{16}, -\frac{1}{2}\right)$ and $\left(\frac{1}{4}, 1\right)$ and hence the required area $= \frac{9}{64}$.

Example 25:

Find the area common to the two curves $y^2 = ax$, $x^2 + y^2 = 4ax$.

Solution:

$y^2 = ax$ is a parabola with vertex at the origin and axis along x-axis and latus rectum a, and $x^2 + y^2 = 4ax$ is a circle with centre (2a, 0) and radius 2a.

Both these curves are symmetrical about x-axis.

Solving the equations of the two curves for x we have

$$x^2 + ax = 4ax$$

or $$x^2 - 3ax = 0$$

or $$x(x - 3a) = 0.$$

Therefore $x = 0, 3a.$

Thus, the two curves intersect at the points where $x = 0$ and $x = 3a$.
Also A is the point (4a, 0).

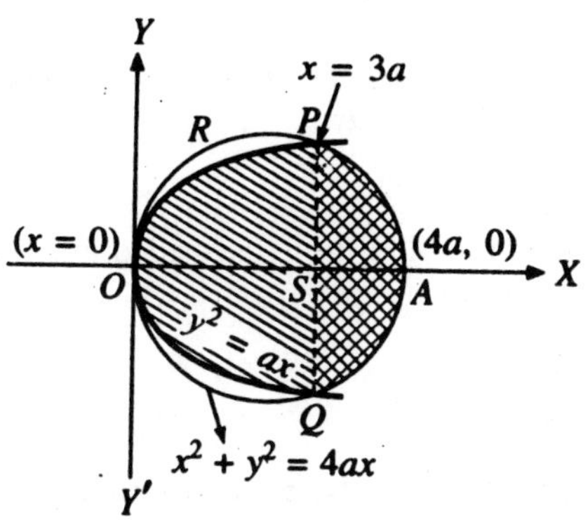

Fig 1.15

The area common to the parabola and the circle (*i.e.*, the shaded area)

$$= 2\ [\text{Area OPS} + \text{Area PSA}],\ (\text{by symmetry})$$

$$= 2\left[\int_0^{3a} y\,dx, \text{ form the parabola } y^2 = ax\right.$$

$$\left. = \int_{3a}^{4a} y\,dx, \text{ from the circle } x^2 + y^2 = 4ax\right]$$

$$= 2\left[\int_0^{3a} \sqrt{(ax)}\,dx + \int_{3a}^{4a} \sqrt{(4ax - x^2)}\,dx\right]$$

$$= 2\sqrt{a}\int_0^{3a} x^{1/2}\,dx + 2\int_{3a}^{4a} \sqrt{[4a^2 - (x - 2a)^2]}\,dx$$

$$= 2\sqrt{a}\left[\frac{2}{3}x^{3/2}\right]_0^{3a}$$

$$+ 2\left[\frac{1}{2}(x - 2a)\sqrt{\{4a^2 - (x - 2a)^2\}} + \frac{4a^2}{3}\sin^{-1}\frac{x - 2a}{2a}\right]_{3a}^{4a}$$

$$= 4a^2\sqrt{3} + 2\left[\left\{0 - \frac{1}{2}a\sqrt{3}.a\right\} + 2a^2\{(\pi/2) - (\pi/6)\}\right]$$

$$= 4\sqrt{3}\,a^2 - a^2\sqrt{3} + \frac{4}{3}\pi a^2$$

$$= 3\sqrt{3a^2} + \frac{4}{3}\pi a^2$$

$$= a^2\left(3\sqrt{3} + \frac{4}{3}\pi\right)$$

Example 26:

Find the area lying above x-axis and included between the circle $x^2 + y^2 = 2ax$ and the parabola $y^2 = ax$.

Solution:

Here we are required to find the is unshaded area ORP.

Solving the given equations $y^2 = ax$

and $x^2 + y^2 = 2ax$ for x,

we get $x = 0$ and $x = a$.

$\therefore$ the required area $= \int_0^a y\,dx$, from the curve $x^2 + y^2 = 2ax$

$- \int_0^a y\,dx$, from the curve $y^2 = ax$

$$= \int_0^a \sqrt{(2ax - x^2)}\,dx - \int_0^a \sqrt{(ax)}\,dx$$

$$= \int_0^a \{a^2 - (x - a)^2\}\,dx - \sqrt{a}.\left[\frac{2}{3}x^{3/2}\right]_0^a$$

$$= \left[\frac{1}{2}(x-a)\sqrt{(2ax - x^2)} + \frac{a^2}{2}\sin^{-1}\left(\frac{x-a}{a}\right)\right]_0^a - \frac{2}{a}a^2$$

$$= \left[0 - \frac{a^2}{2}\sin^{-1}(-1)\right] - \frac{2}{3}a^2 = \frac{a^2}{2}.\frac{\pi}{2} - \frac{2}{3}a^2$$

$$= a^2\left[\frac{\pi}{4} - \frac{2}{3}\right].$$

Example 27:

Prove that the area of the region bounded by the parabolas $y = x^2$ and $y = 4 - x^2$ is $\frac{16}{3}\sqrt{2}$.

Solution:

The given parabolas are

$$y = x^2 \quad ...(1)$$

and

$$y = 4 - x^2 \quad ...(2)$$

Both the curves (1) and (2) are symmetrical about y-axis.

The equation of the parabola (2) can be written as $x^2 = -(y-4)$ which shows that its vertex is the point (0, 4).

Solving (1) and (2) for x,

we get $x^2 = 4 - x^2$

or $2x^2 = 4$ or $x^2 = 2$

or $x = \pm\sqrt{2}$.

Putting $x = \sqrt{2}$ in (1) and (2) meet at the points $(-\sqrt{2}, 2)$j and $(\sqrt{2}, 2)$.

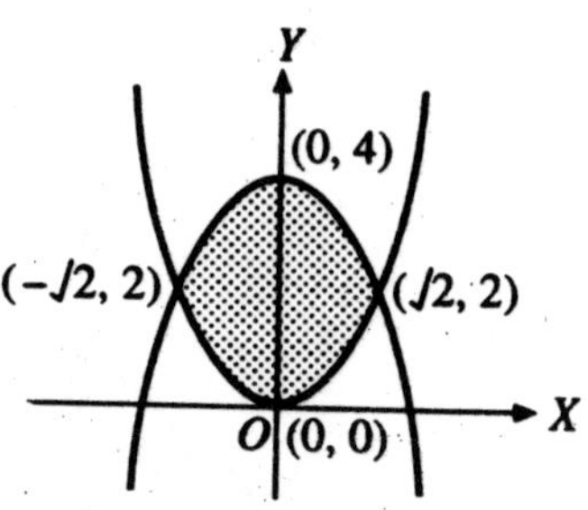

Fig 1.16

Required area bounded by the two parabolas

$$= 2\left[\int_0^{\sqrt{2}} \underset{\text{From}(2)}{y\,dx} - \int_0^{\sqrt{2}} \underset{\text{From}(1)}{y\,dx}\right]$$

$$= 2\left[\int_0^{\sqrt{2}} (4-x^2)dx - \int_0^{\sqrt{2}} x^2 dx\right]$$

$$= \int_0^{\sqrt{2}} 2(4-2x^2)\,dx$$

$$= 2\left[4x - \frac{2}{3}x^3\right]_0^{\sqrt{2}}$$

$$= 2\left[4\sqrt{2} - \frac{2}{3}.2\sqrt{2}\right]$$

$$= 2.4\sqrt{2}\left[1-\frac{1}{3}\right] = 2.4\sqrt{2}\,.\,\frac{2}{3} = \frac{16}{3}\sqrt{2}.$$

Example 28:

Show that the area included between the parabolas $y^2 = 4a(x+a)$, $y^2 = 4b(b-x)$ is $\frac{8}{3}(a+b)\sqrt{(ab)}$.

Solution:

$y^2 = 4a(x+a)$ represents a parabola whose vertex is $(-a, 0)$ and latus rectum is 4a. Also $y^2 = 4b(b-x)$ represents a parabola whose vertex is (b, 0) and latus rectum 4b. Both the curves have been shown in the figure.

Equating the values of y^2 from the two given equations of parabolas, we get $4a(x+a) = 4b(b-x)$ or $x = b - a$ *i.e.*, the abscissa of the point of intersection P is $b - a$.

Now both the curves are symmetrical about x-axis.

∴ the required area = 2 [Area. APM + Area PMB], by symmetry

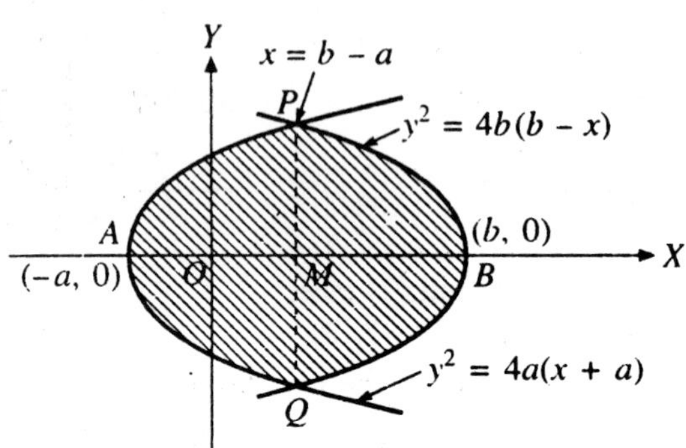

Fig 1.17

$$= 2\left[\int_{-a}^{b-a} y\,dx, \text{ for the parabola } y^2 = 4a(x+a)\right.$$

$$\left. + \int_{b-a}^{b} y\,dx, \text{ for the parabola} y^2 = 4b(b-x)\right]$$

$$= 2\left[\int_{-a}^{b-a} \sqrt{\{4a(x+a)\}}dx + \int_{b-a}^{b} \sqrt{\{4b(b-x)\}}dx\right]$$

$$= 4\sqrt{a}\int_{-a}^{b-a}(x+a)^{1/2}\,dx + 4\sqrt{b}\int_{b-a}^{b}\sqrt{(b-x)}\,dx$$

$$= 4\sqrt{a}\left[\frac{2}{3}(x+a)^{3/2}\right]_{-a}^{b-a} - 4\sqrt{b}\left[\frac{2}{3}(b-x)^{3/2}\right]_{b-a}^{b}$$

$$= \frac{1}{3}[8\sqrt{a}.b^{3/2}] + \frac{1}{3}[8\sqrt{b}.a^{3/2}]$$

$$= \frac{8}{3}\sqrt{(ab)}.(b+a) = \frac{8}{3}(a+b)\sqrt{(ab)}.$$

Example 29:

Show that the area common to the ellipses $a^2x^2 + b^2y^2 = 1$, $b^2x^2 + a^2y^2 = 1$, where $0 < a < b$, is $4\,(ab)^{-1}\tan^{-1}(a/b)$.

Solution:

The given equations of the two ellipses are

$$a^2x^2 + b^2y^2 = 1 \qquad ...(1)$$

and

$$b^2x^2 + a^2y^2 = 1 \qquad ...(2)$$

Since $0 < a < b$, therefore $(1/a) > (1/b)$.

The ellipse (1) cuts the x-axis on the positive side at the point $(1/a, 0)$ and it cuts the y-axis on the positive side at the point B $(0, 1/b)$

The ellipse (2) cuts the x-axis on the positive side at the point A $(1/b, 0)$.

Both the ellipses are symmetric about both the axes.

Solving (1) and (2) we have the coordinates of the point of intersection P in the first quadrant as $1/\sqrt{\{a^2 + b^2\}}$, $1/\sqrt{\{a^2 + b^2\}}$.

Draw PM and PN perpendiculars to the axis of x and the axis of y respectively.

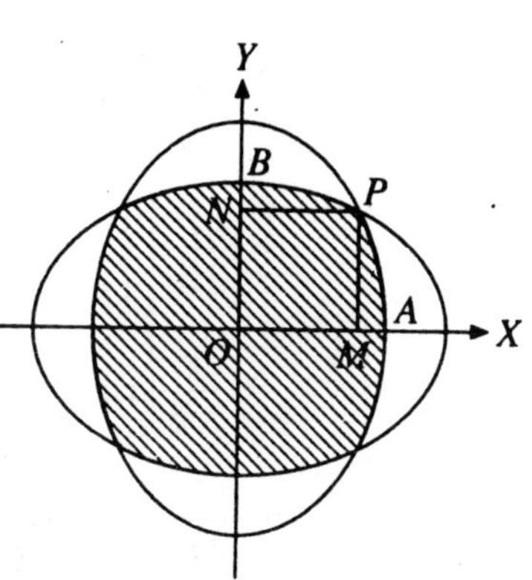

Fig 1.18

Now the area common the two ellipses (*i.e.*, the shaded area)

$= 4 \times$ (common area in the Ist quadrant) $= 4$ area OAPB

$= 4$[area OMPB + area APM]

$= 4$[{area of the square OMPN + area BPN} + area APM]

$= 4$[area of the square OMPN + 2 area APM] ...(3)

[$\therefore$ area BPN = area APM on account of the symmetrical situation of the area OAPB about OP]

Now the area of the square OMPN = OM.ON

$$= \frac{1}{\sqrt{(a^2+b^2)}} \cdot \frac{1}{\sqrt{(a^2+b^2)}} = \frac{1}{(a^2+b^2)}.$$

Also the area APM

$$= \int_{1\sqrt{(a^2+b^2)}}^{1/b} y\, dx, \text{ form the ellipse } b^2x^2 + a^2y^2 = 1$$

$$= \int_{1/\sqrt{(a^2+b^2)}}^{1/b} \frac{(1-b^2x^2)^{1/2}}{a} dx$$

$$= \frac{b}{a} \int_{1/\sqrt{(a^2+b^2)}}^{1/b} \sqrt{\left[\frac{1}{b^2} - x^2\right]} dx$$

$$= \frac{b}{a} \left[\frac{x}{2} \sqrt{\left(\frac{1}{b^2} - x^2\right)} + \frac{1}{2} \cdot \frac{1}{b^2} \sin^{-1}\left(\frac{x}{1/b}\right) \right]_{1/\sqrt{(a^2+b^2}}^{1/b}$$

$$= \frac{b}{2a}\left[0 + \frac{1}{b^2}\sin^{-1} 1 - \frac{1}{\sqrt{(a^2+b^2)}} \cdot \frac{a}{b\sqrt{(a^2+b^2)}} - \frac{1}{b^2}\sin^{-1}\frac{b}{\sqrt{(a^2+b^2)}}\right]$$

$$= \frac{1}{2ab}\left[\frac{\pi}{2} - \frac{ab}{a^2+b^2} - \sin^{-1}\frac{b}{\sqrt{(a^2+b^2)}}\right]$$

$$= \frac{1}{2ab}\left[\left\{\frac{\pi}{2} - \sin^{-1}\frac{b}{\sqrt{(a^2+b^2)}}\right\} - \frac{ab}{a^2+b^2}\right]$$

$$= \frac{1}{2ab}\left[\cos^{-1}\left\{\frac{b}{\sqrt{(a^2+b^2)}}\right\} - \frac{ab}{a^2+b^2}\right]$$

$$= \frac{1}{2ab}\left[\tan^{-1}\left(\frac{a}{b}\right) - \frac{ab}{a^2+b^2}\right].$$

Hence from (3), the required area

$$= 4\left[\frac{1}{a^2+b^2} + 2\left\{\frac{1}{2ab}\tan^{-1}\frac{a}{b} - \frac{1}{2(a^2+b^2}\right\}\right]$$

$$= \frac{4}{ab}\tan^{-1}\left(\frac{a}{b}\right).$$

Example 30:

If P(x, y) be any point on the ellipse $x^2/a^2 + y^2/b^2 = 1$ and S be the sectorial area bounded by the curve, the x-axis and the line joining the origin to P, show that $x = a \cos (2S/ab)$, $y = b \sin (2S/ab)$.

Solution:

We have S = the sectorial area OAP (*i.e.*, the dotted area)

= the area of the Δ OMP + the area PMA

$$= \frac{1}{2}\text{OM.MP} + \int_x^a y\,dx, \text{ for the ellipse}$$

$$= \frac{1}{2}xy + \int_x^a \frac{b}{a}\sqrt{(a^2 - x^2)}\,dx,$$

[∴ from the equation of the ellipse, $y = (b/a)\sqrt{(a^2 - x^2)}$]

$$= \frac{1}{2}x.\frac{b}{a}\sqrt{(a^2 - x^2)} + \frac{b}{a}\left[\frac{x}{2}\sqrt{(a^2x^2)} + \frac{1}{2}a^2\sin^{-1}\frac{x}{a}\right]_x^a$$

$$= \frac{bx}{2a}\sqrt{(a^2 - x^2)}$$

$$+ \frac{b}{a}\left[0 + \frac{1}{2}a^2.\frac{\pi}{2} - \frac{x}{2}\sqrt{(a^2 - x^2)} - \frac{1}{2}a^2 \sin^{-1}\frac{x}{a}\right]$$

$$= \frac{bx}{2a}\sqrt{(a^2 - x^2)} + \frac{b}{a}.\frac{1}{2}a^2\left(\frac{\pi}{2} - \sin^{-1}\frac{x}{a}\right)$$

$$- \frac{bx}{2a}\sqrt{(a^2 - x^2)}$$

$$= \frac{ab}{2}\left(\frac{\pi}{2} - \sin^{-1}\frac{x}{a}\right)$$

$$= \frac{ab}{2}\cos^{-1}\frac{x}{a}.$$

Thus $S = \frac{ab}{2}\cos^{-1}\frac{x}{a}$;

$\therefore$ $\cos^{-1}\frac{x}{a} = \frac{2S}{ab}$

or $\frac{x}{a} = \cos\frac{2S}{ab}$

or $x = a\cos\frac{2S}{ab}$.

Also $y = \frac{b}{a}\sqrt{(a^2 - x^2)}$

$$= \frac{b}{a}\sqrt{\{a^2 - a^2\cos^2(2S/ab)\}}$$

$$= b\sin\frac{2S}{ab}.$$

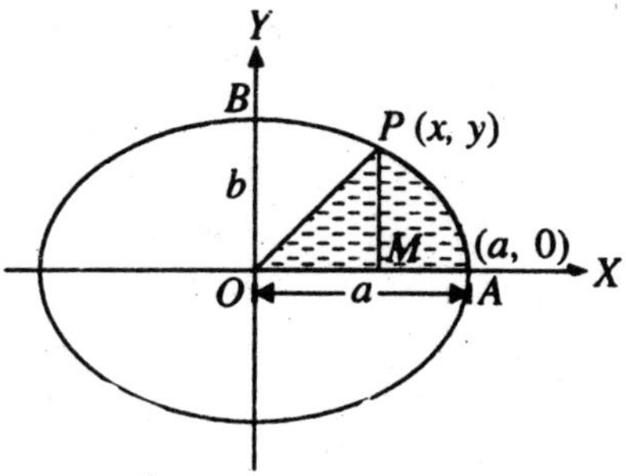

Fig 1.19

Example 31:

If A is the vertex, O the centre and P any point (x, y) on the hyperbola $x^2/a^2 - y^2/b^2 = 1$, show that $x = a\cosh(2S/ab)$, $y = b\sinh(2S/ab)$, where S is the sectorial area OPA.

Solution:

The given hyperbole is shown is the figure. We have S = the sectorial area OAP (*i.e.*, the dotted area)

= the area of the D OMP – the area PAM

$$= \frac{1}{2}OM.MP - \int_a^x y\,dx, \text{ for the hyperbola}$$

$$= \frac{1}{2}xy - \int_a^x \frac{b}{a}\sqrt{(x^2 - a^2)}\,dx,$$

[$\because$ from the equation of the hyperbola $y = (b/a)\sqrt{(x^2 - a^2)}$]

$$= \frac{1}{2}x.\frac{b}{a}\sqrt{(x^2 - a^2)} - \frac{b}{a}\left[\frac{x}{2}\sqrt{(x^2 - a^2)} - \frac{1}{2}a^2 \cosh^{-1}\frac{x}{a}\right]_a^x$$

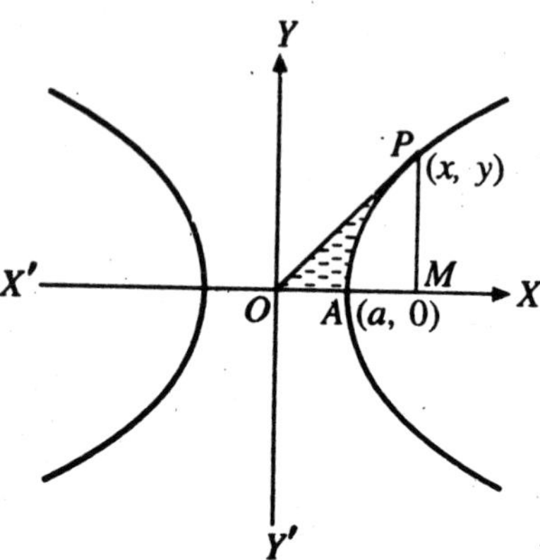

Fig 1.20

$$= \frac{bx}{2a}\sqrt{(x^2 - a^2)}$$

$$- \frac{b}{a}\left[\frac{x}{2}\sqrt{(x^2 - a^2)} - \frac{1}{2}a^2 \cosh^{-1}\frac{x}{a} - (0 - 0)\right]$$

$$= \frac{bx}{2a}\sqrt{(x^2 - a^2)} - \frac{bx}{2a}\sqrt{(x^2 - a^2)} + \frac{ab}{2}\cosh^{-1}\frac{x}{a}\cosh^{-1}\frac{x}{a}.$$

Thus $S = \frac{ab}{2}\text{csoh}^{-1}\frac{x}{a}$;

$\therefore \quad \cos^{-1}\frac{x}{a} = \frac{2S}{ab}$ or $x = a \cosh\frac{2S}{ab}$.

Also $y = \frac{b}{a}\sqrt{(x^2 - a^2)} = \frac{b}{a}\sqrt{\left\{a^2 \cosh^2\frac{2S}{ab} - a^2\right\}}$

$$= \frac{b}{a}.a\sqrt{\left\{\cosh^2\frac{2S}{ab} - 1\right\}} = b \sinh\frac{2S}{ab}.$$

Example 32:

Prove that the area of a sector of the ellipse of semi-axes a and b between the major axis and a radius vector from the focus is 1/2ab) θ – e sin θ), where θ is the eccentric angle of the point to which the radius vector is drawn.

Solution:

Let the equation of the ellipse be $x^2/a^2 + y^2 = 1$. Let O be its centre and S be its focus (ae, 0). Let θ be the eccentric angle of any point P (x, y) on the ellipse. Then x = a cos θ, y = b sin θ

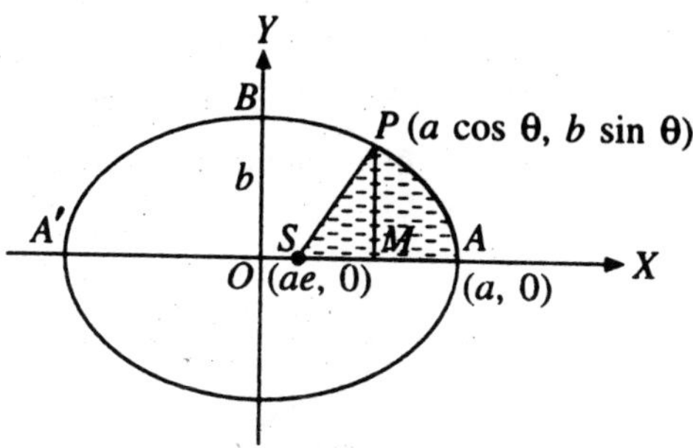

Fig 1.21

Now SP is the radius vector of P drawn through the focus S and SA is the radius vector along the major axis. At the point A, x = a and θ = 0.

Draw PM perpendicular to the x-axis.

$$= \text{area of the } \Delta \text{ SMP} + \text{area PMA}$$

$$= \frac{1}{2}\text{SM.MP} + \int_{a\cos\theta}^{a} y\,dx, \text{ for the ellipse}$$

$$= \frac{1}{2}(\text{OM} - \text{OS}).\ \text{MP} + \int_{\theta}^{0} y\frac{dx}{d\theta}d\theta$$

$$= \frac{1}{2}(a\cos\theta - ae)\, b\sin\theta + \int_{\theta}^{0} b\sin\theta.\,(-a\sin\theta)\,d\theta,$$

$$[\therefore\ x = a\cos\theta \text{ and } y = b\sin\theta]$$

$$= \frac{1}{2}ab\,(\cos\theta - e)\sin\theta + \int_{0}^{\theta} ab.\ \frac{1}{2}(1 - \cos 2\theta)\,d\theta$$

$$= \frac{1}{2}ab\,(\cos\theta - e)\sin\theta + \frac{1}{2}ab\left[\theta - \frac{\sin 2\theta}{2}\right]_0^{\theta}$$

$$= \frac{1}{2}ab\,(\cos\theta - e)\sin\theta + \frac{1}{2}ab\,(\theta - \frac{1}{2}.\ 2\sin\theta\cos\theta)$$

$$= \frac{1}{2}ab\,[\cos\theta\sin\theta - e\sin\theta + \theta - \sin\theta\cos\theta]$$

$$= \frac{1}{2}ab\,(\theta - e\sin\theta).$$

Example 33:

Find the area common to the circle $x^2 + y^2 = 4$ and the ellipse $x^2 + 4y^2 = 9$.

Solution:

The equation of the circle is $x^2 + y^2 = 4$, ...(1)

and the equation of the ellipse is $x^2 + 4y^2 = 9$. ...(2)

Both the curves (1) and (2) are symmetrical about both the axes and have been shown in the figure.

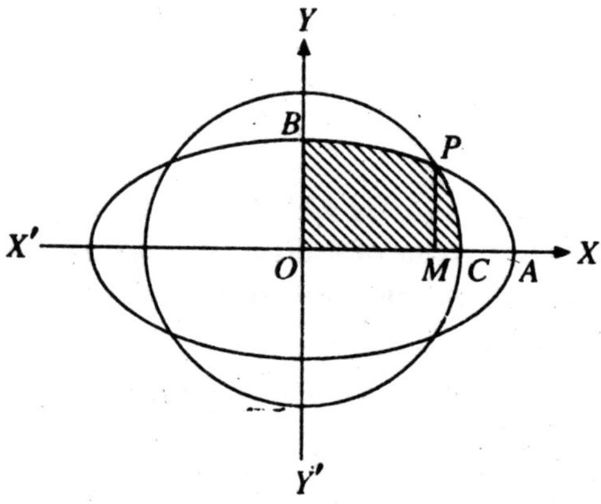

Fig 1.22

Solving (1) and (2) for x, we have $x^2 + 4(4 - x^2) = 9$ or $3x^2 = 7$ or $x^2 = 7/3$.

∴ the x-coordinate of the point of intersection P lying in the first quadrant is $\sqrt{(7/3)}$.

Also putting $y = 0$ in $x^2 + y^2 = 4$,

we get $x = 2$ at C.

Now the required area is symmetrical about both the axes.

∴ the required area (*i.e.*, the area common to the circle and the ellipse)

= 4 × (common area lying in the firs quadrant) = 4 × area OCPB

= 4[area OBPM + area CPM]

$$= 4\left[\int_0^{\sqrt{(7/3)}} y\,dx, \text{ for the ellipse} + \int_{\sqrt{(7/3)}}^{2} y\,dx, \text{ for the circle}\right].$$

$$= 4\left[\int_0^{\sqrt{(7/3)}} \frac{1}{2}\sqrt{(9-x^2)}dx + \int_{\sqrt{(7/3)}}^{2} \sqrt{(4-x^2)}dx\right],$$

[$\therefore$ for the ellipse, $y = \frac{1}{2}\sqrt{(9 - x^2)}$ and for the circle, $y = \sqrt{(4 - x^2)}$]

$$= 2\left[\frac{x\sqrt{(9-x^2)}}{2} + \frac{9}{2}\sin^{-1}\left(\frac{x}{3}\right)\right]_0^{\sqrt{(7/3)}}$$

$$+ 4\left[\frac{x\sqrt{(4-x^2)}}{2} + 2\sin^{-1}\left(\frac{x}{2}\right)\right]_{\sqrt{(7/3)}}^{2}$$

$$= 2\left[\frac{1}{2}\cdot\sqrt{\left(\frac{7}{3}\right)}\sqrt{\left(\frac{20}{3}\right)} + \frac{9}{2}\sin^{-1}\left\{\frac{1}{3}\sqrt{\left(\frac{7}{3}\right)}\right\}\right]$$

$$+ 4\left[2\sin^{-1}(1) - \frac{1}{2}\sqrt{\left(\frac{7}{3}\right)}\sqrt{\left(\frac{5}{3}\right)} - 2\sin^{-1}\left\{\frac{1}{2}\sqrt{\left(\frac{7}{3}\right)}\right\}\right]$$

$$= \frac{2\sqrt{(35)}}{3} + 9\sin^{-1}\frac{\sqrt{7}}{3\sqrt{3}} + 4\pi - \frac{2}{3}\sqrt{(35)} - 8\sin^{-1}\frac{\sqrt{7}}{2\sqrt{3}}$$

$$= 4\pi + 9\sin^{-1}\left\{\frac{1}{3}\cdot\sqrt{(7/3)}\right\} - 8\sin^{-1}\left\{\frac{1}{2}\cdot\sqrt{(7/3)}\right\}$$

Example 34:

Show that the larger of the two areas into which the circle $x^2 + y2 = 64a^2$ is divided by the parabola $y^2 = 12ax$ is $\frac{16}{3}a^2$ [8p – √3].

Solution:

$x^2 + y^2 = 64a^2$ is a circle with centre (0, 0) and radius 8a and $y^2 = 12ax$ is a parabola whose vertex is at (0, 0) and latus rectum 12a. Both the curves are symmetrical about x-axis. Solving the two equations, the co-ordinates of the common point P are (4a, 4a √3). Draw PM perpendicular from P to the y-axis.

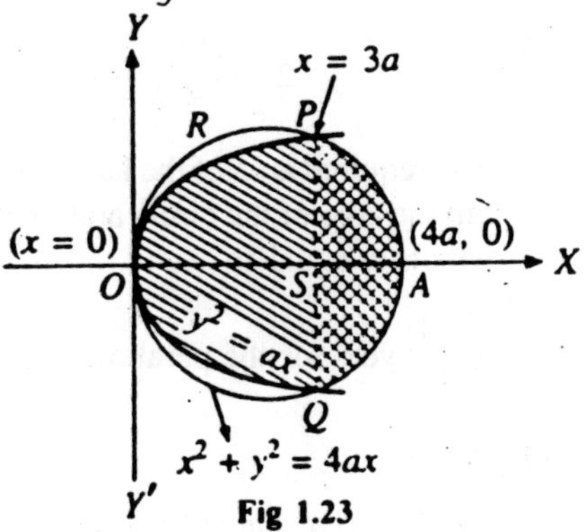

Fig 1.23

Now the area of the circle (*i.e.*, the shaded area) = the area PRSTQOP

= the area of the semi-circle RST + 2 area OPR

$$= \frac{1}{2} \cdot \pi\ (8a)^2 + 2\ [\text{area OPM} + \text{area MPR}]$$

$$= \frac{1}{2}\pi\ (8a)^2 + 2\int_0^{4a\sqrt{3}} x\ dy, \text{ for } y^2 = 12\ ax$$

$$+ 2\int_{4a\sqrt{3}}^{8a} x\ dy, \text{ for } x^2 + y^2 = 64a^2$$

$$= 32\pi\ a^2 + 2\int_0^{4a\sqrt{3}} \frac{y^2}{12a} dy + 2\int_{4a\sqrt{3}}^{8a} \sqrt{(64a^2 - y^2)}\ dy$$

$$= 32\pi a^2 + \frac{1}{6a}\left[\frac{y^3}{3}\right]_0^{4a\sqrt{3}}$$

$$+ 2\left[\frac{1}{2} y \sqrt{(64a^2 - y^2)} + \frac{64a^2}{2} \sin^{-1} \frac{y}{8a}\right]_{4a\sqrt{3}}^{8a}$$

$$= 32\pi a^2 + \frac{1}{6a}\left[\frac{64 \times 3\sqrt{3}a^3}{3}\right]$$

$$+ 2[\{0 - 8a^2\ \sqrt{3}\} + 32a^2\ \{\sin^{-1} 1 - \sin^{-1} (\sqrt{3/2})\}]$$

$$= 32\pi a^2 + \frac{32\sqrt{3}a^2}{3} - 16a^2\ \sqrt{3} + \frac{32}{3} a^2\pi$$

$$= \frac{128}{3} a^2\pi - \frac{16}{3} a^2\ \sqrt{3} = \frac{16}{3} a^2\ (8\pi - \sqrt{3}).$$

Notes:

1. The required area can also be found as follows:
   the required area

$$= 2\left[\{\int_{-8a}^{4a} y dx, \text{for the circle} - \int_0^{4a} y dx, \text{for the parabols}\right].$$

2. The required area of the larger portion can also be evaluated by finding out the area of the circle and then subtracting from it the area APOQA which is equal to

$$2\left[\int_0^{4a} y dx, \text{for the parabola} + \int_{4a}^{8a} y dx, \text{for the circle}\right].$$

Example 35:

Find the area common to the circle $x^2 + y^2 = 9$ and the parabola $x^2 = 8y$.

Solution:

Proceed as in part (a) above.

Example 36:

Find the area included between the parabola $x^2 = 4ay$ and the curve $y = 8a^3/(x^2 + 4a^2)$.

Solution:

The curve $y(x^2 + 4a^2) = 8a^3$ is symmetrical about y-axis. Equating to zero the coefficient of the highest power of x in the given equation, we find that $y = 0$ *i.e.*, x-axis is an asymptote of the curve parallel to x-axis. Also this curve cuts the y-axis at (0, 2a).

Solving the two given equations $x^2 = 4ay$ and $y = 8a^3/(x^2 + 4a^2)$ we get their points of intersection as $(\pm 2a, a)$.

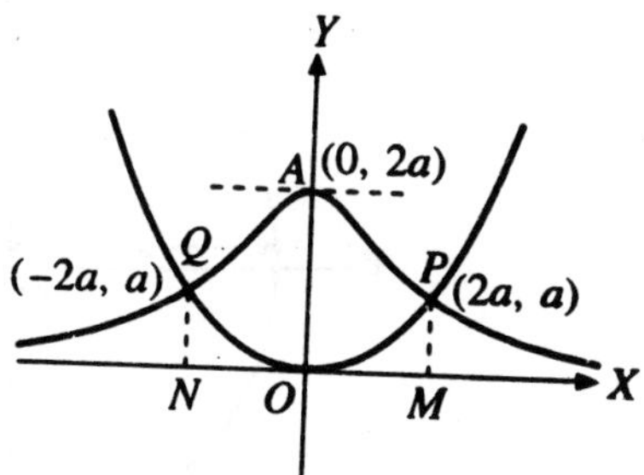

Fig 1.24

Also both the curves are symmetrical about y-axis.

Now the required area OPAQO = 2 × area OPA (by symmetry)

$$= 2 \times [\text{area OAPM} - \text{area OPM}]$$

$$= 2\left[\int_0^{2a} y\,dx, \text{for } y = 8a^3/(x^2 + 4a^2) - \int_0^{2a} y\,dx, \text{for } x^2 = 4ay\right]$$

$$= \int_0^{2a} \frac{8a^3}{x^2 + 4a^2}\,dx - 2\int_0^{2a} \frac{x^2}{4a}\,dx$$

$$= 16a^3.\frac{1}{2a}\left[\tan^{-1}.\frac{x}{2a}\right]_0^{2a} - \frac{1}{2a}\left[\frac{x^3}{3}\right]_0^{2a}$$

$$= 2\pi a^2 - \frac{4a^2}{3}$$

$$= \left[2\pi - \frac{4}{3}\right]a^2.$$

Example 37:

Find by double integration the area of the region enclosed by the curves $x^2 + y^2 = a^2$, $x + y = a$ (in the first quadrant).

Solution:

The given equations of the circle $x^2 + y^2 = a^2$ [centre (0, 0) and radius a] and of the straight line $x + y = a$ (with equal intercepts a on both the axes) can be easily traced as shown in the figure.

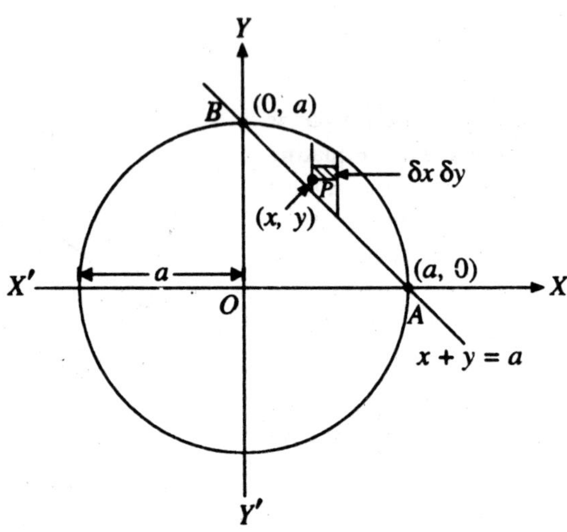

Fig 1.25

The required area is the area bounded by the arc AB and the line AB. To find it with the help of double integration take any point P(x, y) in this portion and consider an elementary area δx δy at P. The required area can now be covered by first moving y from the straight line $x + y = a$ to the arc of the arc of the circle $x^2 + y^2 = a^2$ and then moving x from 0 to a.

∴ the required area

$$= \int_{x=0}^{a}\int_{y=(a-x)}^{\sqrt{(a^2-x^2)}} dx\, dy,$$ the first integration to be performed w.r.t. y whose limits are variable

$$= \int_0^a [y]_{(a-x)}^{\sqrt{(a^2-x^2)}} dx$$

$$= \int_0^a [\sqrt{(a^2 - x^2)} - (a - x)]\, dx$$

$$= \left[\left\{\frac{1}{2}x\sqrt{(a^2 - x^2)} + \frac{1}{2}a^2 \sin^{-1}(x/a)\right\} - ax + \frac{1}{2}x^2\right]_0^a$$

$$= \frac{1}{2}a^2.\left(\frac{1}{2}\pi\right) - a^2 + \frac{1}{2}a^2$$

$$= \frac{1}{2}a^2\left(\frac{1}{2}\pi - 1\right) = \frac{1}{4}a^2\ (\pi - 2).$$

Note : The required area can also be covered by first moving x from the st. line $x + y = a$ to the arc of the circle $x^2 + y^2 = a^2$ and then moving y from 0 to a.

Example 38:

Find by double integration the area bounded by the curves $y(x^2 + 2) = 3x$ *and* $4y = x^2$.

Solution:

Eliminating y from the given equation, we get $3x/(x^2 + 2) = x^2/4$

or $$12x = x^2 (x^2 + 2)$$

or $$x\ (12 - x^3 - 2x) = 0,$$

giving $x = 0$ and $x = 2$.

Thus the two curves intersect at the points where $x = 0$ and $x = 2$.

Both the curves have been shown in the figure.

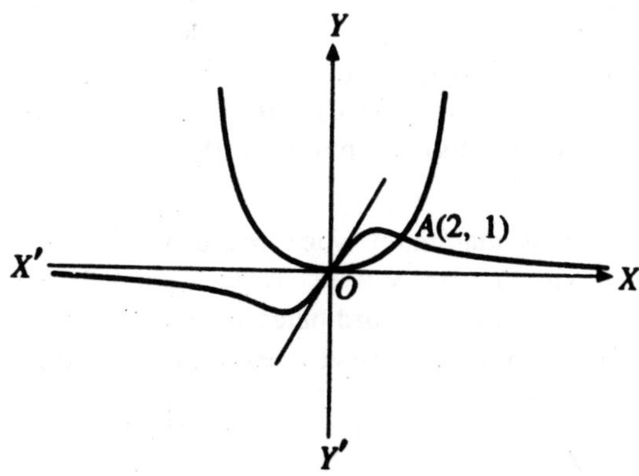

Fig 1.26

The required area

$$= \int_{x=0}^{2}\int_{y=x^2/4}^{3x/(x^2+2)} dx\ dy,$$

the first integration to be performed w.r.t. y

$$= \int_0^2 [y]_{x^2/4}^{3x/(x^2+2)} dx = \int_0^2 \left[\frac{3x}{x^2+2} - \frac{x^2}{4}\right] dx$$

$$= \frac{3}{2}\int_0^2 \frac{2.xdx}{x^2+2} - \frac{1}{2}\int_0^2 x^2\,dx = \frac{3}{2}\left[\log(x^2+2)\right]_0^2 - \frac{1}{4}\left[\frac{x^3}{3}\right]_0^2$$

$$= \frac{3}{2}[\log 6 - \log 2] - \frac{1}{12}.[8] = \frac{3}{2}\log\left(\frac{6}{2}\right) - \frac{2}{3}$$

$$= \frac{3}{2}\log 3 - \frac{3}{2}.$$

Example 39:

Find the area included between the cycloid x = a (θ – sin θ), y = a (1 – cos θ) and its base.

Solution:

The parametric equations of the given cycloid are $x = a(\theta - \sin\theta)$, $y = a(1 - \cos\theta)$.

We have $dx/d\theta = a(1 - \cos\theta)$, $dy/d\theta = a\sin\theta$.

$$\therefore \frac{dy}{dx} = \frac{dy\,d\theta}{dx/d\theta} = \frac{a\sin\theta}{a(1-\cos\theta)} \frac{2\sin\frac{1}{2}\theta\cos\frac{1}{2}\theta}{2\sin^2\frac{1}{2}\theta} = \cot\frac{1}{2}\theta.$$

In this curve $y = 0$ when $a(1 - \cos\theta) = 0$ *i.e.*, $\cos\theta = 1$ *i.e.*, $\theta = 0$ When $\theta = 0$, $x = a(0 - \sin 0) = 0$, $y = 0$ and $dy/dx = \cot 0 = \infty$. Thus the curve passes through the point (0, 0) and the axis of y is tangent at this point.

In this curve y is maximum when $\cos\theta = -1$ *i.e.*, $\theta = \pi$. When $\theta = \pi$, $x = a(\pi - \sin\pi) = a\pi$, $y = 2a$, $dy/dx = \cot 1/2\pi = 0$. Thus at the point $\theta = \pi$, whose cartesian co-ordinates are $(a\pi, 2a)$, the tangent to the curve is parallel to x-axis. This curve does not exist in the region $y > 2a$.

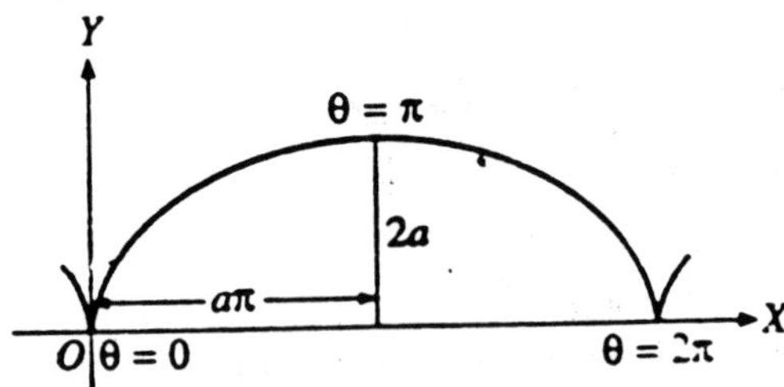

Fig 1.27

In this curve y cannot be –ive because cos θ cannot be greater than

Thus, on complete arch of the give cycloid as shown in the figure.

Now this cycloid is symmetrical with respect to the line $x = a\pi$ (axis of the cycloid) a and its base is the x-axis. Therefore the required area

$$= 2\int_{x=0}^{a\pi} y\,dx = 2\int_{\theta=0}^{\pi} y\frac{dx}{d\theta}.\,d\theta$$

$$= 2\int_0^{\pi} a(1-\cos\theta).\,a(1-\cos\theta)\,d\theta = 2a^2\int_0^{\pi}(1-\cos\theta)^2\,d\theta$$

$$= 2a^2\int_0^{\pi}\left(2\sin^2\frac{1}{2}\theta\right)^2 d\theta = 8a^2\int_0^{\pi}\sin^4 1/2\theta\,d\theta$$

$$= 8a^2\int_0^{\pi/2}\sin^4\phi\,2d\phi, \text{ putting } \frac{1}{2}\theta = \phi \text{ so that } \frac{1}{2}d\theta = d\phi$$

$$= 16a^2\int_0^{\pi/2}\sin^4\phi\,d\phi = 16a^2.\frac{3.1}{4.2}.\frac{\pi}{2}, \text{ by Walli's formula} = 3\pi a^2.$$

Example 40:

Find the area of a loop of the curve $x = a \sin 2t$, $y = a \sin t$ or $a^2x^2 = 4y^2 (a^2 - y^2)$.

Solution:

To trace the given curve, we first find its Cartesian equation by eliminating. We have $x = a \sin 2t = 2a \sin t \cos t$.

$\therefore\ x^2 = 4a^2 \sin^2 t \cos^2 t$

$= 4a^2 \sin^2 t\,(1 - \sin^2 t)$

$= 4a^2 (y^2/a^2)\,\{1 - (y^2/a^2)\}$,

$[\therefore\ y = a \sin t]$

or $a^2x^2 = 4y^2 (a^2 - y^2)$ is the cartesian equation of the given curve.

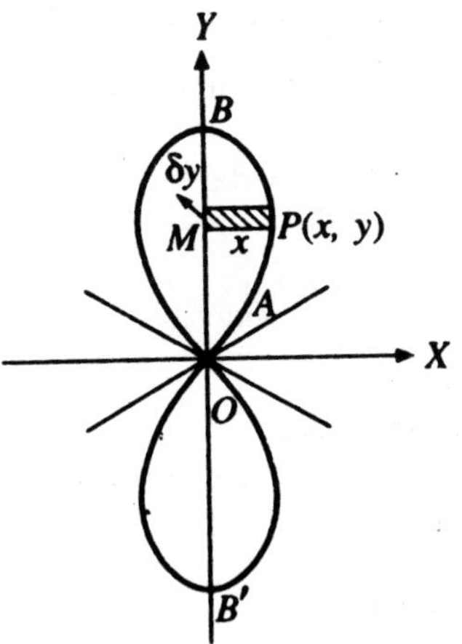

Fig 1.28

Now we trace the curve from its cartesian equation. The curve is as shown in the figure.

At O, $x = 0$, $y = 0$ and so $t = 0$.

Again at B, $x = 0$,

$y = s\,a$ and so $t = \frac{1}{2}\pi$.

The required area of a loop of the curve

$$= 2 \times \text{area OAB} \qquad \text{(By symmetry)}$$

$$= 2\int_{y=0}^{a} x\,dy = 2\int_{t=0}^{\pi/2} x.\frac{dy}{dt}.dt$$

$$= 2\int_{0}^{\pi/2} a \sin 2t.\ a \cos t\ dt$$

$$= 2a^2\int_{0}^{\pi/2} \sin 2t \cos t\ dt = 4a\int_{0}^{\pi/2} \sin t \cos 2t\ dt$$

$$= 4a^2\frac{1}{3.1}.\ 1, \qquad \text{(by Walli's formula)}$$

$$= 4a^2/3.$$

Example 41:

Show that the area bounded by the cissoid $x = a \sin^2 t$, $y = (a \sin^3 t)/\cos t$ and its asymptote is $3\pi a^2/4$.

Solution:

Eliminating t from the given parametric equations, we get

$$y^2 = a^2\frac{\sin^6 t}{\cos^2 t} = \frac{a^2(x^3/a^3)}{1-\sin^2 t} = \frac{x^3/a}{1-(x/a)} = \frac{x^3}{(a-x)}.$$

Therefore the *Cartesian equation of the given curve is $y^2(a-x) = x^3$.*

The curve is symmetrical about x-axis, passes through the origin, the tangent there being $y = 0$. Also there are no real points of the curve if $x < 0$ or if $x > a$. This line $x = a$ is an asymptote of the curve. When $x = 0$, $t = 0$ and when $x = a$, $t = \pi/2$.

$\therefore$ the required are

$$= 2\int_{x=0}^{a} y\,dx = 2\int_{t=0}^{\pi/2} y.\frac{dx}{dt}.\ dt = 2\int_{0}^{\pi/2}\frac{a\sin^3 t}{\cos t}.\ 2a \sin t \cos t\ dt$$

$$= 4a^2\int_{0}^{\pi/2}\sin^4 t\ dt = 4a^2\,\frac{\Gamma\frac{5}{2}\Gamma\frac{1}{2}}{2\Gamma 3} = 4a^2\,\frac{\frac{3}{2}.\frac{1}{2}\sqrt{\pi}.\sqrt{\pi}}{2.2.1} = \frac{3\pi a^2}{4}.$$

Example 42:

Find the area of the loop of the curve $x = a(1-t^2)$, $y = at(1-t^2)$, where $-1 \le t \le 1$.

Solution:

Eliminating t, we have $y^2 = a^2t^2(1-t^2)^2 = x^2t^2$

$$= x^2\{1-(x/a)\} = x^2(a-x)/a.$$

Therefore $ay^2 = x^2 (a - x)$ is the cartesian equation of the given curve. To trace this curve see Ex. 5, page 7.

The required area of the loop

$$= 2\int_{x=0}^{a} y\,dx = 2\int_{t=1}^{0} y.\frac{dx}{dt}.\ dt,$$

$[\therefore$ when $x = 0,\ t = 1$ and when $x = a,\ t = 0]$

$$= 2\int_{1}^{0} at\,(1 - t^2).\ (-\,2at)\ dt = 4a^2\int_{0}^{1} (t^2 - t^4)\ dt$$

$$= 4a^2\left[\frac{1}{3}t^3 - \frac{1}{5}t^3\right]_0^1 = 4a^2\left[\frac{1}{3} - \frac{1}{5}\right] = \frac{8}{15}a^2.$$

Example 43:

Find the whole area of the curve (hypocycloid) given by the equations $x = a\cos^3 t,\ y = b\sin^3 t$.

Solution:

Eliminating t from the given equations the cartesian equation of the curve is

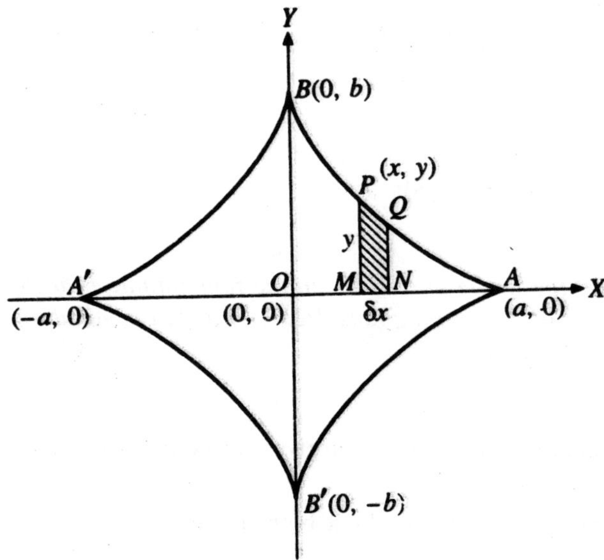

Fig 1.29

$$(x/a)^{2/3} + (y/b)^{2/3} = 1$$

i.e., $$\{(x/a)^2\}^{1/3} + \{(y/b)^2\}^{1/3} = 1.$$

Since the powers of x and y area ll even, the curve is symmetrical about both the axes. It does not pass through the origin. It cuts the axis of x at the points (± a, 0) and the axis of y at the points (0, ± b). The tangent at the point (a, 0) is x-axis. At the point B, x = 0 and t = 1/2π. At the point A, x = a and t = 0.

∴ the required area = 4 × area OAB

$$= 4\int_{x=0}^{a} y\,dx = 4\int_{t=\pi/2}^{0} y.\frac{dx}{at}.\,dt$$

$$= 4\int_{\pi/2}^{0} b\sin^3 t.\,(-3a\cos^2 t\sin t)\,dt,$$

(putting for y and dx/dt)

$$= 12ab\int_{0}^{\pi/2} \sin^4 t\ \text{cso}^2 t\,dt$$

$$= 12ab\,.\frac{3.1.1}{6.4.2}.\frac{\pi}{2} = \frac{3}{8}\pi\,ab.$$

Example 44:

Find the area of the astroid $x^{2/3} + y^{2/3} = a^{2/3}$ *or* $x = a\cos^3 t$, $y = a\sin^3 t$.

Solution:

Do your self. Put b = a.

The required area = $(3/8)\pi\,a^2$.

Example 45:

Prove that the whole area between the four infinite branches of the **tractrix** $x = a\cos t + 1/2a\log\tan 2\ 1/2t$, $y = a\sin t$ *is equal to the area of a circle of radius a.*

Solution:

The parametric equations of the given curve are $x = a\cos t + 1/2a\log\tan^2 1/2t$, $y = a\sin t$.

This curve is symmetrical about both the axes.

We have $$\frac{dx}{dt} = -a\sin t + \frac{1}{2}a\frac{1}{\tan^2\frac{1}{2}t}.\left(2\tan\frac{1}{2}t.\sec^2\frac{1}{2}t\right).\frac{1}{2}$$

$$= -\text{a sin t} + \frac{a}{2\sin\frac{1}{2}t\cos\frac{1}{2}t} = -\text{a sin t} + \frac{a}{\sin t}.$$

$$= \frac{a(1-\sin^2 t)}{\sin t} = \frac{a\cos^2 t}{\sin t}.$$

Also dy/dt = a cos t.

$$\therefore \qquad \frac{dy}{dx} = \frac{dy/dt}{dx/dt} = a\cos t.\frac{\sin t}{a\cos^2 t} = \tan t.$$

In this curve y = 0,

when sin t = 0

i.e., t = 0.

When $t \to 0$, $x \to -\infty$ and $y \to 0$.

Therefore y = 0 is an asymptote of this curve.

In this curve, y is maximum when sin t = 1 *i.e.*, t = 1/2π.

When t = 1/2π, x = 0, y = a, dy/dx = ∞.

Therefore at the point (0, a) the tangent to the curve is perpendicular to the x-axis.

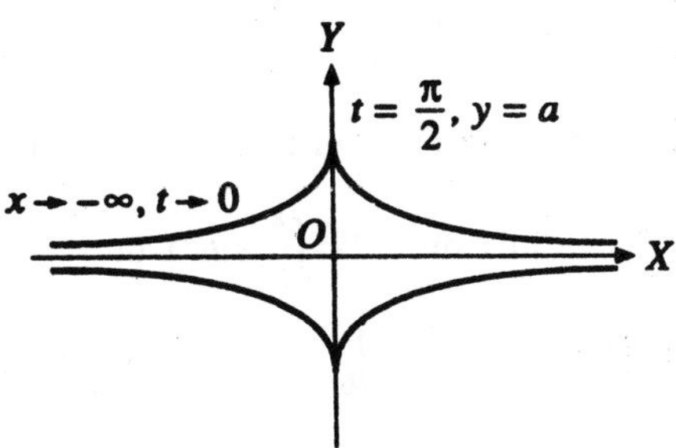

Fig. 1.30

The shape of the curve is as shown in the figure.

The required area = 4 × area lying in the second quadrant

$$= 4\int_{-\infty}^{0} y\,dx = 4\int_0^{\pi/2} y.\frac{dx}{dt}dt = 4\int_0^{\pi/2} a\sin t.\frac{a\cos^2 t}{\sin t}dt$$

$$= 4a^2\int_0^{\pi/2}\cos^2 t\,\delta t = 4a^2.\frac{1}{2}.\frac{\pi}{2}, \quad \text{(by Walli's formula)}$$

$= \pi a^2$ = the area of a circle of radius a.

Example 46:

Find the area included between the curve x= a (t + sin t), y = a(1 – cos t) and its vase.

Solution:

The given cycloid has been shown in the figure.

Since the curve is symmetrical about the y-axis and its base is the line y = 2a, therefore the required area

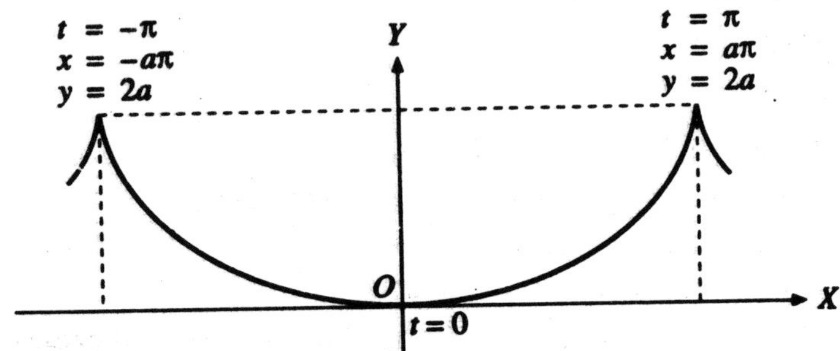

Fig 1.31

$$= 2\int_{y=0}^{2a} x\,dy = 2\int_{t=0}^{\pi} x.\frac{dy}{dt}.\,dt$$

$$= 2\int_0^{\pi} a(t + \sin t)\,a \sin t\,dt = 2a^2 \int_0^{\pi} (t \sin t + \sin^2 t)\,dt$$

$$= 2a^2\int_0^{\pi} t \sin t\,dt + 2a^2\int_0^{\pi} \sin^2 t\,dt$$

$$= 2a^2\left[\{t.(-\cos t\}_0^{\pi} - \int_0^{\pi} 1.(-\cos t)dt\right] + 4a^2\int_0^{\pi/2} \sin^2 t\,dt$$

$$= 2a^2\left[\pi - 0) + \int_0^{\pi} \cos t dt\right] + 4a^2\,1.\frac{1}{2}.\frac{\pi}{2}$$

$$= 2a^2(\pi + 0) + \pi a^2, \left[\because \int_0^{\pi} \cos t dt = 0\right]$$

$$= 3\pi\,a^2.$$

Example 47:

Find the area between the following curves and the given radii vectors:

(i) *The spiral* $rq^{l/2} = a;\ \theta = a,\ \theta = b.$

(ii) *The equiangular spiral* $r = ae^{m\theta};\ \theta = \alpha,\ \theta = \beta.$

(iii) *The parabola $l/r = 1 + \cos\theta$; $\theta = 0$, $\theta = \alpha$.*

Solution:

(i) The required area

$$= \frac{1}{2}\int_{\theta=\alpha}^{\beta} r^2\, d\theta = \frac{1}{2}\int_{\alpha}^{\beta}\frac{a^2}{\theta}d\theta. \qquad (\because r\theta^{1/2} = a)$$

$$= \frac{a^2}{2}\int_{\alpha}^{\beta}\frac{1}{\theta}d\theta = \frac{a^2}{2}\left[\log\theta\right]_{\alpha}^{\beta} = \frac{a^2}{2}\left[\log\beta - \log\alpha\right]$$

$$= \frac{1}{2}a^2 \log(\beta/\alpha).$$

(ii) The required area of the given curve $r = a^{em\theta}$

$$= \frac{1}{2}\int_{\theta=\alpha}^{\beta} r^2\, d\theta = \frac{1}{2}a^2\int_{\alpha}^{\beta} e^{2m\theta}\, d\theta = \frac{1}{2}a^2\left[\frac{e^{2m\theta}}{2m}\right]_{\alpha}^{\beta}$$

$$= (a^2/4m)\,(e^{2m\beta} - e^{2m\alpha}).$$

(iii) The required area

$$= \frac{1}{2}\int_{0}^{\alpha} r^2\, d\theta = \frac{1}{2}\int_{0}^{\alpha}\frac{t^2}{(1+\cos\theta)^2}dx,$$

putting for r from the given equation of the curve

$$= \frac{1}{2}l^2\int_{0}^{\alpha}\frac{d\theta}{\left(2\text{cso}^2\frac{1}{2}\theta\right)^2} = \frac{1}{2}l^2\int_{0}^{\alpha}\frac{d\theta}{4\cos^4\frac{1}{2}\theta}$$

$$= \frac{1}{8}l^2\int_{0}^{\alpha}\sec^4\frac{\theta}{2}d\theta.$$

Now put $\theta/2 = t$

so that $1/2 d\theta = dt$.

Also when $\theta = 0$, $t = 0$

and when $\theta = \alpha$, $t = 1/2\alpha$.

$$\therefore \text{ required area} = \frac{1}{4}l^2\int_{0}^{\alpha/2}\sec^4 t\, dt$$

$$= \frac{1}{4}l^2\int_{0}^{\alpha/2}\sec^2 t.\ \sec^2 t\, dt$$

$$= \frac{1}{4}l^2\int_{0}^{\alpha/2}(1 + \tan^2)\sec^2 t\, dt$$

$$= \frac{1}{4}l^2\int_{0}^{\alpha/2}\{\sec^2 t + (\tan^2 t)\sec^2 t\}dt$$

$$=\frac{1}{4}l^2\left[\tan t+\frac{1}{3}\tan^3 t\right]_0^{\alpha/2},$$

$$\left[\therefore f(\tan t(^2 \sec^2 t dt = \frac{1}{3}(\tan t)^3\right]$$

$$=\frac{1}{4}l^2l^2\left[\tan\frac{1}{2}\alpha+\frac{1}{3}\tan^3\frac{1}{2}\alpha\right].$$

Example 48:

Find the area of the loop of the curve r = a θ cos θ between θ = 0 and θ = π/2.

Solution:

The required area

$$=\frac{1}{2}\int_0^{\pi/2} r^2\,d\theta=\frac{1}{2}\int_0^{\pi/2} a^2\,\theta^2\,\text{cso}^2\theta\,d\theta$$

$$=\frac{a^2}{2}\int_0^{\pi/2}\theta^2\cos^2\theta\,d\theta=\frac{a^2}{4}\int_0^{\pi/2}\theta^2\,(1+\cos 2\theta)\,d\theta$$

$$=\frac{a^2}{4}\int_0^{\pi/2}\theta^2\,d\theta+\frac{a^2}{4}\int_0^{\pi/2}\theta^2\cos 2\theta\,d\theta$$

$$=\frac{a^2}{4}\left[\frac{\theta^3}{3}\right]_0^{\pi/2}+\frac{a^2}{4}\left[\theta^2\,\frac{\sin 2\theta}{2}\right]_0^{\pi/2}-\frac{a^2}{4}\int_0^{\pi/2}2\theta\,.\frac{\sin 2\theta}{2}d\theta$$

$$=\frac{a^2}{12}.\frac{\pi^3}{8}+\frac{a^2}{8}.0-\frac{a^2}{4}\left[\frac{\theta(-\cos 2\theta}{2}\right]_0^{\pi/2}$$

$$+\frac{a^2}{4}\int_0^{\pi/2}1\,.\left[\frac{-\cos 2\theta}{2}\right]d\theta$$

$$=\frac{\pi^3a^2}{96}+\frac{a^2}{8}\left[\frac{\pi}{2}(-1)\right]-\frac{a^2}{8}\left[\frac{\sin 2\theta}{2}\right]_0^{\pi/2}$$

$$=\frac{\pi^3a^2}{96}-\frac{\pi a^2}{16}=\frac{\pi a^2}{96}(\pi^2-6).$$

Example 49:

Find the area of one loop of r = a cos 4θ.

Solution:

The given curve is r = a cos 4θ.

It is symmetrical about the initial line.

One loop is obtained by two consecutive values of θ for which r is zero. We have $r = 0$ when $\cos 4\theta = 0$ *i.e.*, $4\theta = -1/2\pi, 1/2\pi$ *i.e.*, $\theta = -\pi/8, \pi/8$. Thus two consecutive values of θ for which r is zero are $-\pi/8$ and $\pi/8$. Therefore on loop of the curve lies between $\theta = -\pi/8$ and $\pi/8$ and this loop is symmetrical about the initial line $\theta = 0$.

Hence the area of a loop $= 2\int_0^{\pi/8} \frac{1}{2} r^2 \, d\theta$

$$= \int_0^{\pi/8} a^2 \cos^2 4\theta \, d\theta, \qquad [\because r = a \cos 4\theta]$$

$$= \frac{1}{4} a^2 \int_0^{\pi/2} \cos^2 t \, dt, \text{ putting } 4\theta = t \text{ so that } 4\,d\theta = dt$$

$$= \frac{1}{4} a^2 \cdot \frac{1}{2} \cdot \frac{1}{2} \pi, \text{ by Walli's formula}$$

$$= \frac{1}{16} \pi a^2.$$

Example 50:

Find the whole area of the curve $r^2 = a^2 \cos^2\theta + b^2 \sin^2\theta$.

Solution:

The given curve is symmetrical about the initial line ($\because$ the equation of the curve remains unchanged when θ is changed into $-\theta$), and the curve is symmetrical about the line $\theta = \pi/2$ (*i.e.*, y-axis) as the equation remains unchanged when θ is changed into $(\pi - \theta)$. Also there is symmetry about the pole as the equation of the equation of the curve remains unchanged when r is changed into $-$ r.

In this curve r cannot be zero and r is real and finite for all values of θ. The figure of this curve is roughly like that of an ellipse.

$\therefore$ whole area of the curve $= 4 \times$ area lying in the first quadrant

$$= 4 \times \frac{1}{2} \int_{\theta=0}^{\pi/2} r^2 \, d\theta = 2\int_0^{\pi/2} (a^2 \cos^2 \theta + b^2 \sin^2 \theta) \, d\theta$$

$$= 2a^2 \int_0^{\pi/2} \cos^2 \theta \, d\theta + 2b^2 \int_0^{\pi/2} \sin^2 \theta \, d\theta$$

$$= 2a^2 \cdot \frac{1}{2} \cdot \frac{1}{2} \pi + 2b^2 \frac{1}{2} \cdot \frac{1}{2} \pi = \frac{1}{2} \pi (a^2 + b^2).$$

Example 51:

Find the area lying between the cardioid $r = a(1 - \cos\theta)$ *and its double tangent.*

Solution:

Let PQ be the double tangent of the cardioid. Clearly it is perpendicular to OX *i.e.*, it must be inclined at an angle of 90° to the initial line *i.e.*, $\psi = 90°$ at P.

Also we know that at any point of a curve,

$$\psi = \theta + \phi. \qquad ...(1)$$

Now $\tan\phi = r\,(d\theta/dr) = r/(dr/d\theta)$

$$= a\,(1 - \cos\theta)/(a\sin\theta), \qquad [\because r = a(1-\cos\theta)]$$

$$= \frac{2\sin^2\frac{1}{2}\theta}{2\sin\frac{1}{2}\theta\cos\frac{1}{2}\theta} = \tan\frac{1}{2}\theta.$$

$$\therefore \qquad \phi = \frac{1}{2}\theta.$$

Putting the value of ϕ in (1),

we get $\psi = \theta + \frac{1}{2}\theta = \frac{3}{2}\theta.$

Since at P, $\psi = \frac{1}{2}\pi,$

therefore at P, $\frac{1}{2}\theta$ or $\theta = \pi/3.$

$\therefore$ the vectorial angle of the point contact P of the double tangent is $\pi/3$ *i.e.*, 60°. Substituting this value of θ in the equation of the curve, we get the radius vector $OP = a\,(1 - \cos 60°) = a/2.$

Thus in the triangle OPM,

$$OP = \frac{1}{2}a,\ \angle POM = 60°,\ \angle PMO = 90°$$

$$\therefore\ OM = \frac{1}{2}a\cos 60° = \frac{1}{2}a\,.\frac{1}{2}\text{ and PM}$$

$$= \frac{1}{2}a\sin 60° = \frac{1}{2}a\,(\sqrt{3}/2).$$

$\therefore$ area of the triangle OPM $= \frac{1}{2}OM.PM = \frac{1}{2}\left(\frac{1}{4}a\right)(\sqrt{3}a/4) = (1/32)$ $a^2\sqrt{3}.$

Also the sectorial area OPO of the cardioid $r = a\,(1 - \cos\theta)$ *i.e.*, the dotted area

$$= \frac{1}{2}\int_0^{\pi/3} r^2\,d\theta = \frac{1}{2}\int_0^{\pi/3} a^2\,(1 - \cos\theta)^2\,d\theta$$

$$= \frac{1}{2}a^2 \int_0^{\pi/3}\left(1-2\cos\theta+\frac{1}{2}+\frac{1}{2}\cos 2\theta\right)d\theta$$

$$= \frac{1}{2}a^2\left[\frac{3}{2}\theta - 2\sin\theta + \frac{1}{4}\sin 2\theta\right]_0^{\pi/3}$$

$$= \frac{1}{2}a^2\left(\frac{1}{2}\pi - \sqrt{3} + \frac{1}{8}\sqrt{3}\right) = \frac{1}{16}a^2\,(4\pi - 7\sqrt{3}).$$

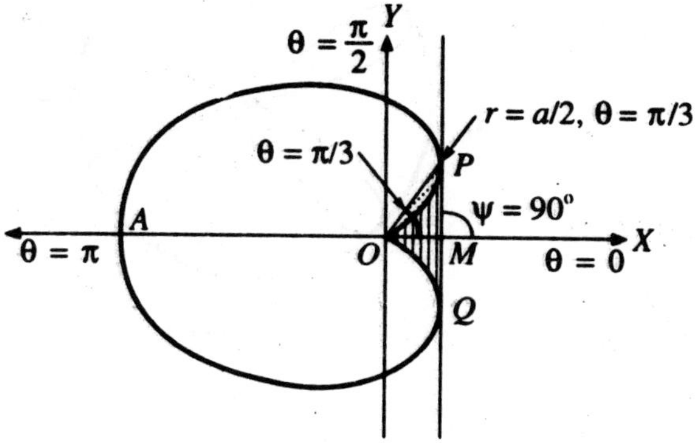

$\theta = \frac{\pi}{2}$ Y

$r = a/2,\ \theta = \pi/3$

$\theta = \pi/3$ P

$\psi = 90^\circ$

A

$\theta = \pi$ O M $\theta = 0$ X

Q

Fig. : 1.32

Hence the required area (*i.e.*, the area shaded by vertical lines = 2 [area of Δ OPM – area of sector OPO]

$$= 2\left[\frac{1}{32}a^2\sqrt{3} - \frac{1}{16}a^2(4\pi - 7\sqrt{3})\right]$$

$$= \frac{1}{16}a^2\,(15\sqrt{3} - 8\pi).$$

Example 52:

Show that the area contained between the circle r = a and the curve r = a cos 5θ is equal to three-fourth of the area of the circle.

Solution:

Curve $r = a\cos 5\theta$ has five loops ($\because$ here n = 5 is odd).

Also putting r = 0, we get $\cos 5\theta = 0$

$$\text{or } 5\theta = \pm\frac{1}{2}\pi \text{ or } \theta = \pm\pi/10.$$

Therefore θ varies from $-\pi/10$ to $\pi/10$ for, the first loop. Also the curve is symmetrical about the initial line. Further giving values to θ from 0 to 2π in $r = a\cos 5\theta$, we observe that the maximum value of r is a and hence the curve r = a cos 5θ lies completely inside the circle r = a as shown in the figure.

Now area of the five loops of the curve r = a cos 5θ

$= 5$ times area of one loop

$$= 5\int_{-\pi/10}^{\pi/10} \frac{1}{2} r^2 \, d\theta$$

$$= \frac{5}{2}\int_{-\pi/10}^{\pi/10} a^2 \cos^2 5\theta \, d\theta$$

$$= \frac{5}{2} \cdot 2\int_{0}^{\pi/10} a^2 \cos^2 5\theta \, d\theta$$

$$= a^2\int_{0}^{\pi/2} \cos^2 \phi \, d\phi,$$

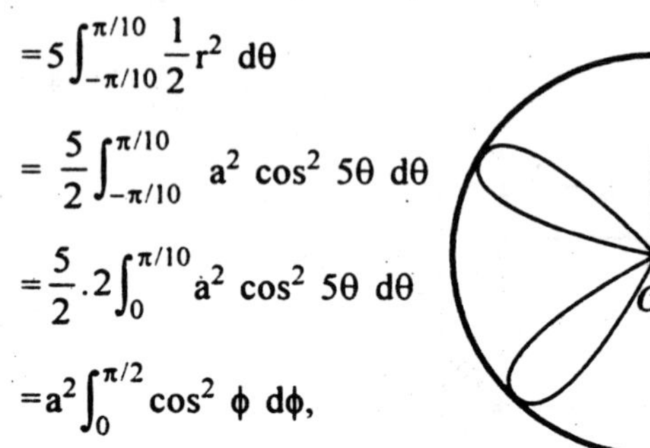

Fig 1.33

putting 5θ = φ

so that 5 dθ = dφ;

also when θ = 0, φ = 0

and when θ = π/10, φ = π/2

$$= a^2 \cdot \frac{1}{2} \cdot \frac{1}{2} p = \frac{1}{4} \pi a^2.$$

Also area of the circle = πa^2.

∴ area contained between the two curves

= area of the

= (3/4) of the area of the circle

Example 53:

Find the area between the curve r = a (sec θ + cos θ) and its asymptote.

Solution:

The given curve is symmetrical about the initial line. When θ = 0, r = 2a.

The given equation of the curve can be written as r = a {(1/cos θ) + cos θ} = a (1 + cos2θ)/cos θ

or $$\frac{1}{r} = \frac{\cos\theta}{a(1+\cos^2\theta)} = f(\theta), \text{ say.}$$

Now f (θ) = 0 gives cos θ = 0 *i.e.*, θ = π/2.

Also $f'(\theta) = \dfrac{1}{a}\dfrac{-\sin\theta(1+\cos^2\theta)-\cos\theta(-2\sin\theta\cos\theta)}{(1+\cos^2\theta)^2}$ ...(1)

Putting $\theta = \pi/2$ in (1),
we get $f'(\pi/2) = -1/a$.

$\therefore$ asymptote of the curve is $r \sin\left(\theta - \frac{1}{2}\pi\right) = 1/(-1/a)$

or $r \cos\theta = a$ *i.e.*, $r = a \sec\theta$.

The cartesian equation of the asymptote is $x = a$.

To find the area between the curve and its asymptote.

Let OQP be a radius vector cutting the curve at P and the asymptote $r = a \sec\theta$ at Q. Let $\angle XOP = \theta$.

Then the shaded area MAPQM = area OAPO – area OMQO.

Now area OAPO $= \int_0^\theta \frac{1}{2} r^2 \, d\theta$, for the given curve

$$= \frac{1}{2}\int_0^\theta a^2 (\sec\theta + \cos\theta)^2 \, d\theta$$

$$= \frac{1}{2}\int_0^\theta a^2 (\sec^2\theta + \cos^2\theta + 2) \, d\theta.$$

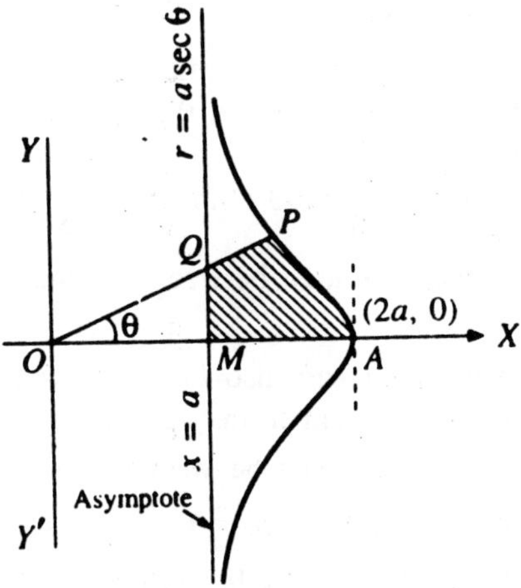

Fig. 1.34

Also area OMQO $= \int_0^{\theta} \frac{1}{2} r^2 \, d\theta$, for the st. line $r = a \sec \theta$

$$= \frac{1}{2} \int_0^{\theta} a^2 \sec^2\theta \, d\theta.$$

$\therefore$ the shaded area MAPQM

$$= \frac{1}{2} \int_0^{\theta} a^2 (\sec^2\theta + \cos^2\theta + 2) \, d\theta - \frac{1}{2} \int_0^{\theta} a^2 \sec^2\theta \, d\theta$$

$$= \frac{1}{2} \int_0^{\theta} a^2 (\cos^2\theta + 2) \, d\theta$$

$$= \frac{1}{2} a^2 \int_0^{\theta} \left\{ \frac{1}{2}(1 + \cos 2\theta) + 2 \right\} d\theta$$

$$= \frac{1}{2} a^2 \int_0^{\theta} \left(\frac{5}{2} + \frac{1}{2} \cos 2\theta \right) d\theta$$

$$= \frac{1}{2} a^2 \left[\frac{5}{2}\theta + \frac{1}{2} \frac{\sin 2\theta}{2} \right]_0^{\theta}$$

$$= \frac{1}{2} a^2 \left(\frac{5}{2}\theta \frac{1}{4} \sin 2\theta \right) = \frac{1}{8} a^2 (10\,\theta + \sin 2\theta).$$

Now, we move the point P further off along curve. Its vectorial angle goes on increasing and ultimately when its distance from O tends to infinity its vectorial angle tends to $\pi/2$.

Hence, the area between the curve and its asymptote lying above the x-axis = limit of the shaded area when $\theta \to \pi/2$

$$= \lim_{\theta \to \pi/2} \frac{a^2}{8} [10\theta + \sin 2\theta] = \frac{a^2}{8} \cdot 10 \cdot \left(\frac{1}{2}\pi \right) = \frac{5}{8} \pi \, a^2.$$

$\therefore$ by symmetry, the total area between the curve and the asymptote $= 2 \cdot (5\pi \, a^2/8) = 5\theta \, a^2/4$.

Aliter : The above area can also be easily found by changing the equation of the curve to cartesian from.

The equation of the curve can be written as

$$r \cos \theta = a\,(1 + \cos^2 \theta) \text{ or } r \cos \theta - a = a \cos^2 \theta$$

or $\quad r^2(r \cos \theta - a) = a\, r^2 \cos^2 \theta$, multiplying both sides by r^2.

Now putting $r \cos \theta = x$ and $r^2 = x^2 + y^2$, we get

$$(x^2 + y^2)(x - a) = ax^2 \text{ or } y^2 (x - a) = ax^2 - x^2 (x - a)$$

or $y^2 (x - a) = x^2 (2a - x)$, which is the cartesian equation of the given curve.

No., trace the curve with the help of this cartesian equation. The curve si symmetrical about x-axis. It meets the x-axis at the point (2a, 0) and the line x = a is an asymptote of the curve. Here origin is a conjugate point because we get imaginary tangents at the origin. The curve does not exist in the regions x > 2a and x <a. It exists only in the region a < x ≤ 2a.

Hence the required area

$$= 2\int_0^{2a} y\,dx = 2\int_0^{2a} x\sqrt{\left(\frac{2a - x}{x - a}\right)}dx$$

(putting for y from the given equation of the curve).

Now put a + a $\sin^2\theta$ so that dx = 2a sin θ cos θ dθ.

Also when x = a, sin θ =

or 0 or θ = 0

and when = 2a,

sin θ = 1 or θ = π/2.

∴ the required area

$$= 2\int_0^{\pi/2} (a + a\sin^2\theta)\frac{\cos\theta}{\sin\theta}\cdot 2a\sin\cos\theta\,d\theta$$

$$= 4a^2\int_0^{\pi/2} (1 + \sin^2\theta)\cos^2\theta\,d\theta$$

$$= 4a^2\int_0^{\pi/2} (\cos^2\theta + \cos^2\theta\sin^2\theta)d\theta$$

$$= 4a^2\left[\frac{1}{2}\cdot\frac{\pi}{2} + \frac{1.1}{4.2}\cdot\frac{\pi}{2}\right], \text{ by Walli's formula}$$

$$= 4a^2\cdot\frac{1}{2}\cdot\frac{\pi}{2}\left[1 + \frac{1}{4}\right] = 4a^2\cdot\frac{1}{4}\pi\cdot\frac{5}{4} = \frac{5}{4}\pi\,a^2.$$

Example 54:

Find the area of the cardioid r = a (1 + cos θ).

Solution:

The given curve is symmetrical about the initial line since ist equation remains unaltered when θ is changed into – θ.

We have r = 0, when cos θ = – 1 *i.e.*, θ = π. Therefore, the line θ = π is tangent at the pole to the curve. Also r is maximum when cos θ = 1 *i.e.*, θ = 0 and then r = 2a.

When θ increases from 0 to r, r decreases from 2a to 0. thus the curve is as shown in the figure.

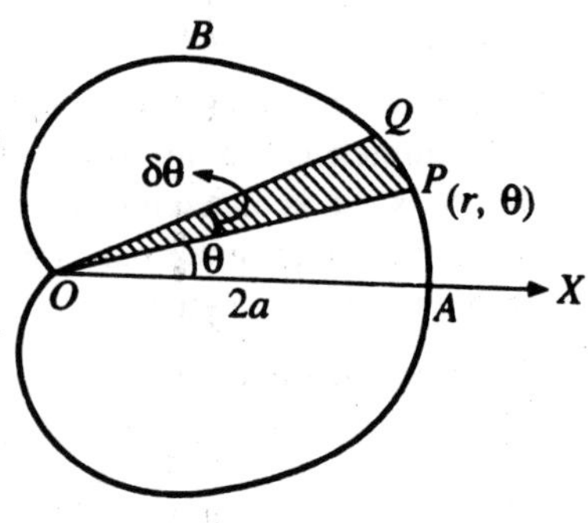

Fig 1.35

Now the required area = 2 × area of the upper half of the curve

$$= 2\int_0^\pi \frac{1}{2} r^2 \, d\theta$$

$$= 2\int_0^\pi \frac{1}{2} a^2 (1 + \cos\theta)^2 \, d\theta,$$

$$[\because r = a(1 + \cos\theta)]$$

$$= a^2 \int_0^\pi \left(2\cos^2 \frac{1}{2}\theta\right)^2 d\theta$$

$$= 4a^2 \int_0^\pi \cos^4 \frac{1}{4}\theta \, d\theta.$$

Now put $1/2\theta = \phi$ so that $\frac{1}{2} d\theta = d\phi$.

Also when $\theta = 0$, $\phi = 0$ and when $\theta = \pi$, $\phi = \pi/2$.

$$\therefore \text{ the required area} = 8a^2 \int_0^{\pi/2} \cos^4 f \, d\theta.$$

$$= 8a^2 . \frac{3.1}{4.2} . \frac{\pi}{2}, \text{ (by Walli's formula} = 3\pi a^2/2)$$

Example 55:

Find the area of the cardioid $r = a(1 - \cos\theta)$.

Solution:

The given curve is symmetrical about the initial line. We have r = 0, when $\cos\theta = 1$ *i.e.*, $\theta = 0$. Therefore the line $\theta = 0$ is tangent at the pole to the curve. Also r is maximum when $\cos\theta = -1$ *i.e.*, $\theta = \pi$ and then r = 2a. When θ increases from 0 to π, r increases from 0 to 2a. Thus the curve is as shown in the figure.

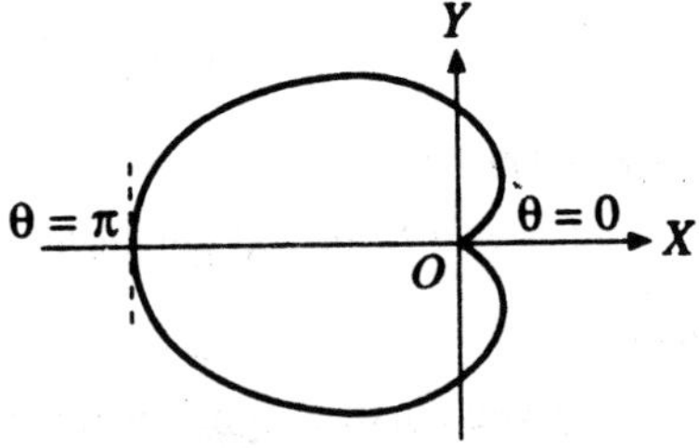

Fig 1.36

The required area

$$= 2\int_0^\pi \frac{1}{2} r^2 \, d\theta = \int_0^\pi a^2 (1 - \cos\theta)^2 \, d\theta$$

$$= a^2\int_0^{\pi}\left(2\sin^2\frac{1}{2}\theta\right)^2 d\theta = 4a^2\int_0^{\pi}\sin^4\frac{1}{4}\theta\, d\theta$$

$$= 8a^2\int_0^{\pi/2}\sin^4\phi\, d\phi.$$

$$\left[\text{putting}\ \frac{1}{2}\theta = \phi\ \text{so that}\ \frac{1}{2}d\theta = d\phi\ \text{and adusting the limits}\right]$$

$$= 8a^2\cdot\frac{3.1}{4.2}\cdot\frac{\pi}{2} = \frac{3\pi a^2}{2}.$$

Example 56:

Show that the area of the limacon $r = a + b\cos\theta$, $(b < a)$ is equal to $\pi\left(a^2 + \frac{1}{2}b^2\right)$.

Solution:

The given curve is symmetrical about the initial line.

We have $r = 0$, when $\cos\theta = -a/b$ *i.e.*, $\theta = \cos^{-1}(-a/b)$ which is not real because $a > b$. Thus, in this curve r cannot be zero Also r is maximum when $\cos\theta = 1$ i.e, $\theta = 0$ and then $r = a + b$. Again r is minimum when $\cos\theta = -1$ *i.e.*, $\theta = \pi$ and then $r = a - b$ which is positive. Some values of r and θ area as follows:

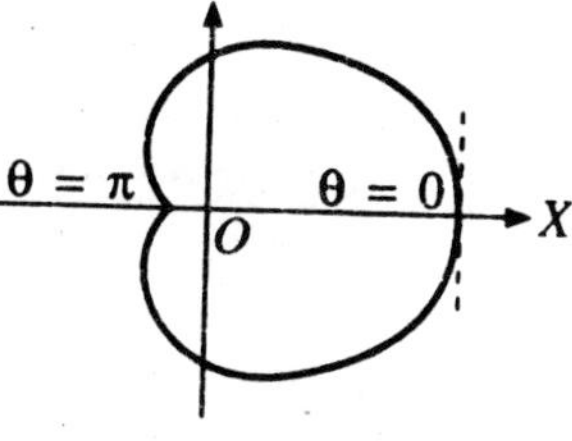

Fig 1.37

θ	0	$\frac{1}{3}\pi$	$\frac{1}{2}\pi$	$\frac{2}{3}\pi$	π
r	$a + b$	$a + \frac{1}{2}b$	a	$a - \frac{1}{2}b$	$a - b$

Hence the curve is as shown in the figure.

$\therefore$ the required area = 2 × area above the initial line

$$= 2\int_0^{\pi}\frac{1}{2}r^2\, d\theta = \int_0^{\pi}(a + b\cos\theta)^2\, d\theta,$$

putting for r from the given equation of the curve

$$= \int_0^{\pi}(a^2 + 2ab\cos\theta + b^2\cos^2\theta)\, d\theta$$

$$= a^2[\theta]_0^{\pi} + 2ab[\sin\theta]_0^{\pi} + \frac{b^2}{2}\int_0^{\pi}(1 + \cos 2\theta)\, d\theta$$

$$= a^2\pi + 0 + \frac{b^2}{2}\left[\theta + \frac{\sin 2\theta}{2}\right]_0^{\pi}$$

$$= a^2\pi + \frac{b^2}{2}\pi = \pi\left(a^2 + \frac{1}{2}b^2\right).$$

Note: If we put b = a or – a, we get the area of the cardioid r = a (1 + cosθ) or r = a (1 – cos θ).

Example 57:

Prove that the sum of the area of the two loops of the limacon $r = a + b\cos\theta$, $(b > a)$ is equal to $\pi(2a^2 + b^2)/2$.

Solution:

The given curve is symmetrical about the initial line. We have r = 0, when cos θ = – a/b *i.e.*, $\theta = \cos^{-1}(-a/b) = \alpha$, (say). Since cos α is –ive, therefore 1/2π < α < π.

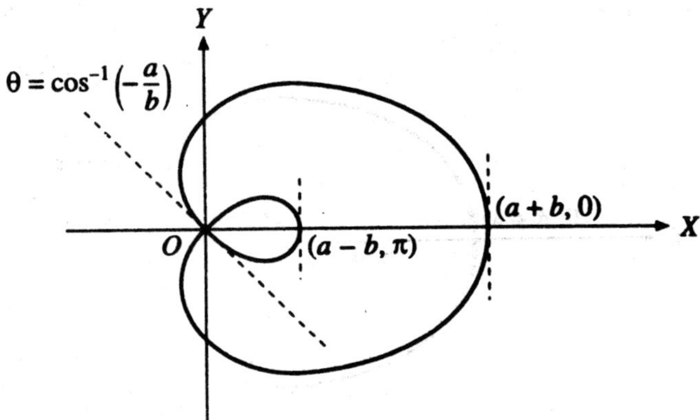

Fig 1.38

Now r is maximum when cos θ = 1 *i.e.*, θ = 0 and then r = a + b.

Also r is minimum when cos θ = – 1 *i.e.*, θ = π and then r = a – b which is negative because b < a. Some values of r and θ are as follows:

θ	0	1/3π	1/2π	α	α < θ < π	π
r	a + b	a + 1/2b	a	0	–ive	a – b

Thus, the curve is as shown in the figure. For the upper half of the larger loop θ varies from 0 to a *i.e.*, $\cos^{-1}(-a/b)$ and for the lower half of the smaller loop θ varies from a to π.

∴ required sum of the two loops

$$= 2\left[\int_0^{\alpha} \frac{1}{2} r^2 d\theta + \int_0^{\alpha} \frac{1}{2} r^2 d\theta\right]$$

$$= 2\int_0^{\pi} \frac{1}{2} r^2 \, d\theta, \text{ by a property o definite integrals}$$

$$= \int_0^{\pi} r^2 \, d\theta = \int_0^{\pi} (a + b \cos \theta)^2 \, d\theta$$

$$= \int_0^{\pi} (a^2 + 2ab \cos \theta + b^2 \cos^2 \theta) \, d\theta$$

$$= \int_0^{\pi} a^2 \, d\theta + 2ab \int_0^{\pi} \cos \theta \, d\theta + b^2 \cdot \int_0^{\pi} \cos^2 \theta \, d\theta$$

$$= a^2[\theta]_0^{\pi} + 0 + 2b^2 \int_0^{\pi/2} \cos^2 \theta \, d\theta$$

$$= a^2 \pi + 2b^2 \cdot \frac{1}{2} \cdot \frac{1}{2} \pi = a^2 \pi + \frac{1}{2} b^2 \pi$$

$$= \frac{1}{2} \pi \, (2a^2 + b^2).$$

Example 58:

Calculate the ratio of the area of the larger to the area of the smaller loop of the curve r = 1/2 + cos 2θ.

Solution:

The given curve is symmetrical about the initial line. In the given equation of the curve $r = 1/2 + \cos 2\theta$ putting $r = 0$, we get $\cos 2\theta = -1/2$ *i.e.*, $2\theta = \pm 2\theta/3$ or $\pm 4\pi/3$ *i.e.*, $\theta = \pm \pi/3$ or $\pm 2\pi/3$.

The greatest radius vector of the loop lying between $\theta = -1/3\pi$ and $\theta = 1/3\pi$ is given by $\theta = 0$ and it is equation to 3/2. The greatest radius vector of the loop lying between $\theta = 1/3\pi$ and $\theta = 2/3\pi$ is given by $\theta = 1/2\pi$ and its numerical values is 1/2.

Thus, we ovserve that the larger loop lies between $\theta = -\pi/3$ and $\theta = \pi/3$ and it is symmetrical about the initial line $\theta = 0$.

Also the smaller loop lies between $\theta = \pi/3$ and $\theta = 2\pi/3$.

Hence Area of the Larger Loop

$$= 2\int_0^{\pi/3} \frac{1}{2} r^2 \, d\theta = \int_0^{\pi/3} \left(\frac{1}{2} + \text{cso} 2\theta\right)^2 d\theta,$$

putting for r from the given equation of the curve

$$= \int_0^{\pi/3} \left(\frac{1}{4} + \cos 2\theta + \cos^2 2\theta\right) d\theta$$

$$= \int_0^{\pi/3}\left[\frac{1}{4}+\cos 2\theta+\frac{1}{2}(1+\cos 4\theta)\right] d\theta$$

$$= \left[\frac{3}{4}\theta+\frac{1}{2}\sin 2\theta+\frac{1}{2}\frac{\sin 4\theta}{4}\right]_0^{\pi/3}$$

$$= \left[\frac{\pi}{4}+\frac{\sqrt{3}}{4}-\frac{1}{2}\frac{\sqrt{3}}{8}\right] = \frac{1}{4}\left[\pi+\frac{3\sqrt{3}}{4}\right]$$

$$= \frac{1}{16}(4\pi + 3\sqrt{3}).$$

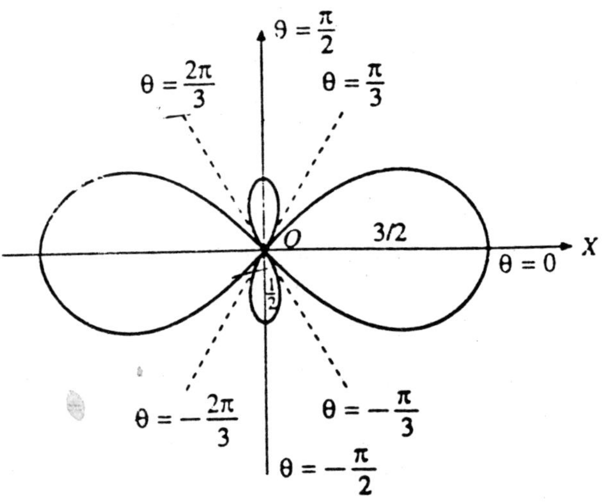

Fig 1.39

Area of the Smaller Loop

(lying between $\theta = \pi/3$ and $\theta = 2\pi/3$)

$$= \int_{\pi/3}^{2\pi/3}\frac{1}{2}r^2\, d\theta = \frac{1}{2}\int_{\pi/3}^{2\pi/3}\left(\frac{1}{2}+\cos 2\theta\right)^2 d\theta$$

$$= \frac{1}{2}\left[\frac{3}{4}\theta+\frac{1}{2}\sin 2\theta+\frac{1}{2}\cdot\frac{\sin 4\theta}{4}\right]_{\pi/3}^{2\pi/3}$$

as in the 1st case

$$= \frac{1}{2}\left[\frac{\pi}{4}+\frac{1}{2}\left(-\frac{\sqrt{3}}{2}-\frac{\sqrt{3}}{2}\right)+\frac{1}{8}\left\{\frac{\sqrt{3}}{2}-\left(-\frac{\sqrt{3}}{2}\right)\right\}\right]$$

$$= \frac{1}{2}\left[\frac{\pi}{4} - \frac{\sqrt{3}}{2} + \frac{\sqrt{3}}{8}\right] = \frac{1}{2}\left[\frac{\pi}{4} - \frac{3\sqrt{3}}{8}\right]$$

$$= \frac{(2\pi - 3\sqrt{3})}{16}.$$

$$\therefore \text{ the required ratio } = \frac{\text{Area of the larger loop}}{\text{Area of the smaller loop}}$$

$$= \frac{\frac{1}{16}(4\pi + 3\sqrt{3})}{\frac{1}{16}(2\pi - 3\sqrt{3})} = \frac{4\pi + 3\sqrt{3}}{2\pi - 3\sqrt{3}}.$$

Example 59:

Show that the area of a loop of $r = a \cos n\theta$ *is* $\pi a^2/4n$, *n being integral. Also prove that the whole area is* $\pi a^2/4$ *or* $\pi a^2/2$ *according as n is odd or even.*

Solution:

The number of loops in $r = a \cos n\theta$ will be n or 2n according ad n is odd or even.

The given curve is symmetrical about the initial line.

Also putting $r = 0$, we have $\cos n\theta = 0$ *i.e.*, $n\theta = -1/2\pi$, $1/2\pi$ *i.e*, $\theta = -\pi/2n, \pi/2n$. Thus two consecutive values of θ for which r is zero are $-\pi/2n$ and $\pi/2n$.

$\therefore$ one loop lies between $\theta = -\pi/2n$ and $\theta = \pi/2n$ and it is symmetrical about the initial line $\theta = 0$.

$$\therefore \text{ area of one loop } = 2.\int_0^{\pi/2n} \frac{1}{2} r^2 \, d\theta, \qquad \text{(By symmetry)}$$

$$= \int_0^{\pi/2n} r^2 \, d\theta = \int_0^{\pi/2n} a^2 \cos^2 n\theta \, d\theta.$$

Now put $n\theta = t$

so that $n \, d\theta = dt$.

Also when $\theta = 0$, $t = 0$

and when $\theta = \pi/2n$, $t = \pi/2$.

$\therefore$ the area of one loop

$$= \frac{a^2}{n}\int_0^{\pi/2} \cos^2 t \, dt = \frac{a^2}{n}.\frac{1}{2}.\frac{\pi}{2} = \frac{\pi a^2}{4n}.$$

Now if n is odd, the total number of loops will be n and the whole area = n × of one loop = n $(\pi a^2/4n)$ = $1/4\pi\ a^2$.

If n is even, the total number of loops will be 2n and then the whole area = 2n × area of one loop = 2n × $(\pi a^2/4n)$ = $1/2\pi\ a^2$.

Example 60:

Find the area of a loop of the curve $r = a \cos 3\theta + b \sin 3\theta$.

Solution:

In the given equation of the curve put a = k cos a, b = k sin a

so that $k = \sqrt{(a^2 + b^2)}$ and a = tan–1 (b/a).

Thus the given equation reduces to

$$r = k \cos 3\theta \cos \alpha + k \sin 3\theta \sin \alpha$$

or

$$r = k \cos (3\theta - \alpha) = k \cos 3\left(\theta - \frac{1}{3}\alpha\right).$$

Now rotating the initial line through an angle $\alpha/3$, the given equation of the curve becomes

$$r = k \cos 3\left(\theta + \frac{1}{3}\alpha - \frac{1}{3}\alpha\right) = k \cos 3\theta.$$

It should be noted that the rotation of the initial line changes only the equation of the curve and has no effect on its shape. Therefore the area of a loop of the given curve is the same as the area of a loop of the curve $r = k \cos 3\theta$.

Thus curve $r = k \cos 3\theta$ is symmetrical about the initial line.

Putting r = 0 in it we have, $\cos 3\theta = 0$ *i.e.*, $3\theta = \pm \pi/2$ *i.e.*,$\theta = \pm \pi/6$.

∴ one loop of this curve lies between $\theta = -\pi/6$ and $\pi = +\pi/6$ and it is symmetrical about the initial line.

∴ the required area $= 2.\int_0^{\pi/6} \frac{1}{2} r^2\, d\theta,$ (By symmetry)

$$= \int_0^{\pi/6} k^2 \cos^2 3\theta\, d\theta.$$

Now put $3\theta = t$,

so that $3\, d\theta = dt$.

Also when $\theta = 0$, $t = 0$

and when $\theta = \pi/6$, $t = \pi/2$.

∴ the required area

$$= \frac{k^2}{3}\int_0^{\pi/2} \cos^2 t \, dt = \frac{k^2}{3}.\frac{1}{2}.\frac{1}{2}\pi = \frac{k^2}{12}\pi$$

$= (a^2 + b^2)\ \pi/12$

$(\therefore\ k^2 = a^2 + b^2).$

Example 61:

Trace the curve $r = \sqrt{3}\cos 3\theta + \sin 3\theta$, *and find the area of a loop.*

Solution:

The given equation of the curve is $r = \sqrt{}\ \cos 3\theta + \sin 3\theta$

$$= 2\left\{\left(\sqrt{3}/2\right)\cos 3\theta + \frac{1}{2}\sin 3\theta\right\},$$

$= 2\{\sin \pi/3 \cos 3\theta + \text{cso}\ \pi/3 \sin 3\theta),$

$$\left[\therefore \sin\pi/3 = \sqrt{3}/2 \text{ and } \cos\pi/3 = \frac{1}{2}\right]$$

$= 2\sin(3\theta + \pi/3) = 2\sin 3(\theta + \pi/9).$

Now turning the initial line through an angle $-\pi/9$ the given equation of the curve becomes $r = 2\sin 3(\theta - \pi/9 + \pi/9) = 2\sin 3\theta$.

Now we shall trace the curve $r = 2\sin 3\theta$.

This curve $r = 2\sin 3\theta$ will have 3 loops and is not symmetrical about the initial line. Also putting $r = 0$, we get $\sin 3\theta = 0$ *i.e.*, $3\theta = 0, \pi$ i.e, $\theta = 0, \pi/3$.

Therefore, the lines $\theta = 0$ and $\theta = \pi/3$ are tangents to the curve at the pole and one loop of this curve lies between these two lines. For this loop r is greatest when $\sin 3\theta = 1$ *i.e.*, $3\theta = \pi/2$ *i.e.*, $\theta = \pi/6$. Thus, this loop bends at $\theta = \pi/6$ and there r is equal to 2.

Here one loop of the curve lies in the region $0 < \theta < \pi/3$. one loop lies in the region $\pi/3 < \theta < 2\pi/3$ and one loop lies in the region $2\pi/3 < \theta < \pi$. If θ increases beyond π to 2π the same branches of the curve are repeated and we get any new branch.

Here $a = 2$.

Now to get the location of the given curve turn the initial line back to its original position *i.e.*, turn the initial line through an angle $\pi/9$.

The required area or a loop

$$= 2 \times \int_0^{\pi/6} \frac{1}{2} r^2 \, d\theta$$

$$= \int_0^{\pi/6} 4\sin^2 3\theta \, d\theta$$

$$= 2\int_0^{\pi/6} (1 - \cos 6\theta)\, d\theta$$

$$= 2\left[\theta - \frac{\sin 6\theta}{6}\right]_0^{\pi/6} = \frac{\pi}{3}.$$

Example 62:

Find the area common to the circles $r = a\sqrt{2}$ and $r = 2a \cos \theta$.

Solution:

The given equations of circles area $r = a\sqrt{2}$ and $r = 2a \cos \theta$. The first equation represents a circle with centre at pole and radius $a\sqrt{2}$. The second equation represents a circle passing through the pole and the diameter through the pole as the initial line. Both these circle area symmetrical about the initial line. Eliminating r between the two equations, we gave at the points of intersection $a\sqrt{2} = 2a \cos \theta$, *i.e.*, $\cos \theta = 1/\sqrt{2}$, *i.e.*, $\theta = \pm\, \pi/4$.

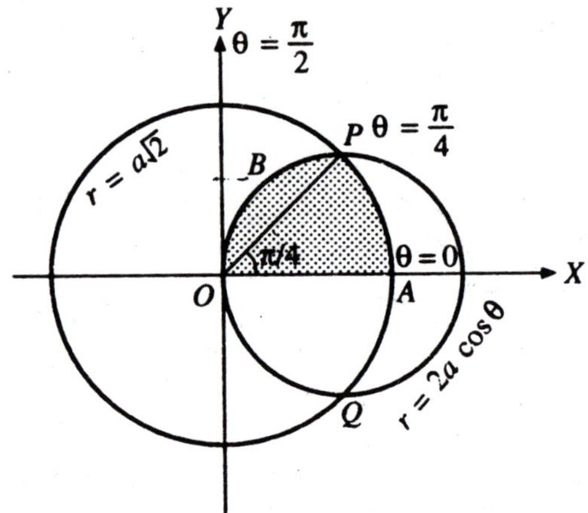

Fig 1.40

Thus at P, $\theta = \pi/4$.

For the circle $r = 2a \cos \theta$,

at O, $r = 0$ and so $\cos \theta = 0$ *i.e.*, $\theta = \frac{1}{2}\pi$.

Now the required area = Area OQAPBO

= 2 (area OAPBO), (by symmetry)

= 2 [Area OAP + Area OPBO]

$$= 2\left[\frac{1}{2}\int_0^{\pi/4} r^2 d\theta, \text{for the circler} = a\sqrt{2}\right.$$

$$\left.+\frac{1}{2}\int_{\pi/4}^{\pi/2} r^2 d\theta, \text{for the circle } r = 2a\cos\theta\right]$$

$$= \int_0^{\pi/4} (a\sqrt{2})^2\, d\theta + \int_{\pi/4}^{\pi/2} (2a \text{ cso } \theta)^2\, d\theta$$

Example 63:

Find the area outside the circle $r = 2a \cos \theta$ and inside the cardioid $r = a (1 + \cos \theta)$.

***Solution*:**

The two curves intersect where $2a \cos \theta = a (1 + \cos \theta)$ *i.e.*, $\cos \theta = 1$ *i.e.*, $\theta = 0$. Besides this the two curves also intersect at the pole $r = 0$. Since for all values of θ, $2a \cos \theta$ *i.e.*, $a \cos \theta + a \text{ cso } \theta \leq a (1 + \cos \theta)$, therefore the circle lies entirely within the cardioid.

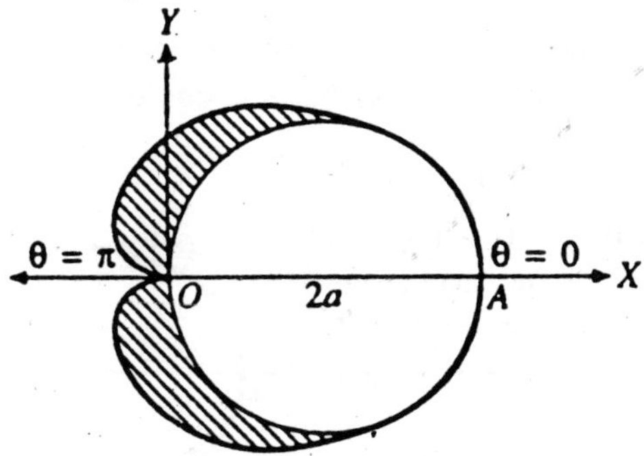

Fig 1.41

Hence the required area

= Area of the cardioid – Area of the circle ...(1)

Now area of the cardioid

$$= 2\int_0^{\pi} \frac{1}{2} r^2\, d\theta$$

$$= \int_0^{\pi} a^2 (1 + \cos \theta)^2\, d\theta$$

$$= a^2 \int_0^{\pi} 4 \cos^4 \frac{1}{2}\theta\, d\theta$$

$$= 8a^2 \int_0^{\pi/2} \cos^4 \phi \, d\theta,$$

$$\left(\text{putting } \frac{1}{2}\theta = \phi \text{ so that } \frac{1}{2}d\theta = d\phi, \text{also when } \theta = 0,\right.$$

$$\left. \phi = 0 \text{ and when } \theta = \pi, \phi = \frac{1}{2}\pi\right)$$

$$= 8a^2 \frac{\Gamma\frac{5}{2}\Gamma\frac{1}{2}}{2\Gamma 3} = 8a^2 \cdot \frac{\frac{3}{2}\cdot\frac{1}{2}\sqrt{\pi}.\sqrt{\pi}}{2.2.1} = \frac{3\pi a^2}{2}$$

And area of the circle = πa^2, because the radius of the circle is a.

$\therefore$ Form (1), the required area = $\frac{3}{2}\pi a^2 - \pi a^2 = \frac{1}{2}\pi a^2$.

Example 64:

Find the total area inside r = sin θ and outside r = 1 – cos θ.

Solution:

Eliminating r between the given equations, we have $\sin\theta = 1 - \cos\theta$, or $\sin^2\theta = (1 - \cos\theta)^2 = 1 + \cos^2\theta - 2\cos\theta$,

or $1 - \cos^2\theta = 1 + \cos^2\theta - 2\cos\theta$, or $2\cos\theta(\cos\theta - 1) = 0$.

$\therefore$ $\cos\theta = 0$ or $\cos\theta = 1$ *i.e.*, $\theta = 0$ or $\pi/2$. Thus, the two curves intersect at the points where $\theta = 0$ and $\theta = \pi/2$.

Draw the two curves in the same figure. The first curve is a circle passing through the pole and the diameter through the line $\theta = \pi/2$. The second curve is a cardioid.

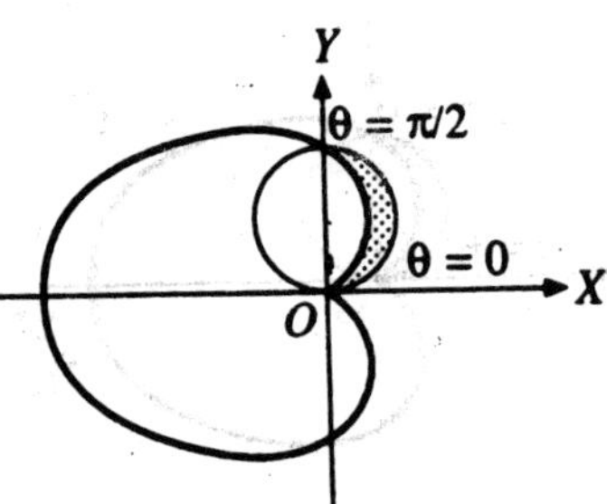

Fig 1.42

$$\therefore \text{ required area} = \frac{1}{2}\int_0^{\pi/2} r^2 \, d\theta, \text{ for the curve } r = \sin\theta$$

$$- \frac{1}{2}\int_0^{\pi/2} r^2 \, d\theta, \text{ for the curve } r = 1 - \cos\theta$$

$$= \frac{1}{2}\int_0^{\pi/2} \sin^2\theta \, d\theta - \frac{1}{2}\int_0^{\pi/2} (1 - \cos\theta)^2 \, d\theta$$

$$= \frac{1}{2}\int_0^{\pi/2} \{\sin^2\theta - (1 - \cos\theta)^2\} \, d\theta$$

$$= \frac{1}{2}\int_0^{\pi/2} 2\ \sin^2\theta\ \cos\ \theta - \cos^2\theta)\ d\theta$$

$$= [\sin\theta]_0^{\pi/2} - \frac{1}{2}.\frac{\pi}{2} = \left(1 - \frac{\pi}{4}\right).$$

Example 65:

Find the area common to the circle r = a and the cardioid r = a (1 + cos θ).

Solution:

Eliminating r between the given equations, we get a (1 + cos θ) = a or cos θ = 0 or θ = ± π/2.

Thus, the two curves cut each other at the point where θ = ± π/2.

Both the curves are symmetrical about the initial line and have been shown in the same figure.

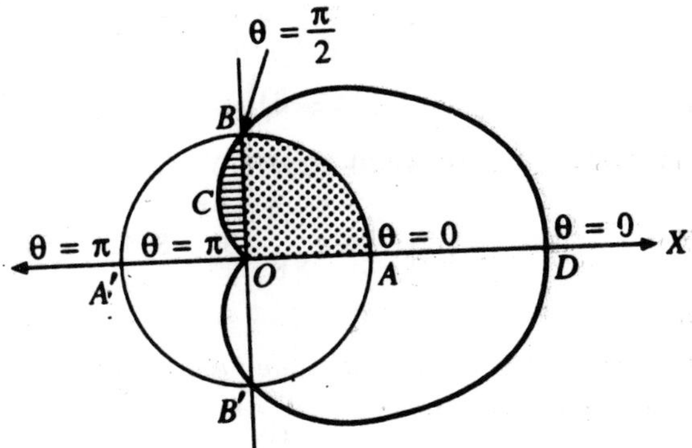

Fig 1.43

Hence the required area = 2 × area ABCOA, [By symmetry)

= 2 × (Area OABO + Area OBCO) ...(1)

Now area OABO $= \frac{1}{2}\int_{\theta=0}^{\pi/2} r^2\ d\theta$, for r = a

$$= \frac{1}{2}a^2\int_0^{\pi/2} d\theta = \frac{1}{2}a^2[\theta]_0^{\pi/2}$$

$$= \frac{1}{2}a^2.\ \frac{\pi}{2} = \frac{\pi a^2}{4}$$

And area OBCO $= \frac{1}{2}\int_{\pi/2}^{\pi} r^2\, d\theta$, for $= a\,(1 + \cos\theta)$

$$= \frac{1}{2}\int_{\pi/2}^{\pi} a^2\,(1 + \cos\theta)^2\, d\theta$$

$$= \frac{1}{2}a^2\int_{\pi/2}^{\pi} (1 + 2\cos\theta + \cos^2\theta)\, d\theta$$

$$= \frac{1}{2}a^2\int_{\pi/2}^{\pi} \left\{1 + \cos\theta + \frac{1}{2}(1 + \cos 2\theta))\right\} d\theta$$

$$= \frac{1}{2}a^2\int_{\pi/2}^{\pi} \left(\frac{3}{2} + 2\cos\theta + \frac{1}{2}(1 + \cos 2\theta\right) d\theta$$

$$= \frac{a^2}{2}\left[\frac{3\theta}{2} + 2\sin\theta + \frac{\sin 2\theta}{4}\right]_{\pi/2}^{\pi}$$

$$= \frac{a^2}{2}\left(\frac{3\pi}{2} - \frac{3\pi}{4} - 2\right)$$

$$= \frac{a^2}{8}(3\pi - 8).$$

Hence form (1), the required area

$$= 2\left[\frac{\pi a^2}{4} + \frac{a^2}{8}(3\pi - 8)\right] = a^2\left(\frac{5\pi}{4} - 2\right).$$

Example 66:

Find the area of the portion included between the cardioids $r = a\,(1 + \cos\theta)$ and $r = a\,(1 - cso\,\theta)$.

Solution:

Eliminating r between the given equations, we get $a\,(1 + \cos\theta) = a\,(1 - \cos\theta)$ or $2\cos\theta = 0$ *i.e.*, $\theta = \pm\,\pi/2$, showing that the two curves meet at the points P $(\theta = \pi/2)$ and Q $(\theta = -\,\theta/2)$. Also by symmetry it is clear that the required area = 4 × area OLPO, where OLP is an arc of the cardioid $r = a\,(1 - \cos\theta)$

$$= 4\int_0^{\pi/2} \frac{1}{2} r^2\, d\theta = 4\int_0^{\pi/2} \frac{1}{2} a^2\,(1 - cso\,\theta)^2\, d\theta$$

$$= 2a^2\int_0^{\pi/2} (1 - 2\cos\theta + \cos^2\theta)\, d\theta$$

$$= 2a^2\int_0^{\pi/2} \left[1 - 2\cos\theta + \frac{1}{2}(1 + \cos 2\theta)\right] d\theta$$

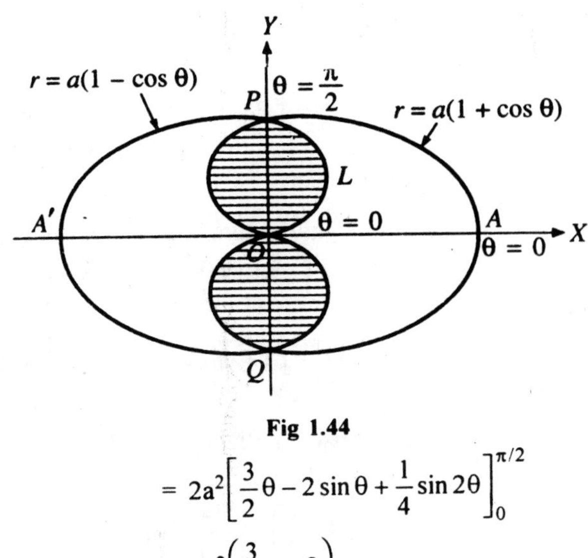

Fig 1.44

$$= 2a^2\left[\frac{3}{2}\theta - 2\sin\theta + \frac{1}{4}\sin 2\theta\right]_0^{\pi/2}$$

$$= 2a^2\left(\frac{3}{4}\pi - 2\right).$$

Example 67:

Find the ratio of the two parts into which the parabola $2a = r(1 + \cos\theta)$ divides the area of the cardioid $r = 2a\ (1 + cso\ \theta)$.

Solution:

Eliminating r between the given equation of the curves, we get

$$2a\ (1 + \cos\theta) = 2a/(1 + \cos\theta) \text{ or } (1 + \cos\theta)^2 = 1$$

or $\cos\theta\ (\cos\theta + 2) = 0$ or $\cos\theta = 0$, $[\because \cos\theta \neq -2]$

or $\theta = \pm\ \pi/2$. Thus at the point of intersection P of the two curves, $\theta = \pi/2$.

Now area of the whole cardioid

$$= 2 \times \frac{1}{2}\int_0^{\pi} r^2\, d\theta, \text{ (by symmetry)}$$

$$= \int_0^{\pi} 4a^2\ (1 + cso\ \theta)^2\, d\theta$$

$$= 4a^2\int_0^{\pi}\left(2\cos^2\frac{1}{2}\theta\right)^2 d\theta = 16a^2\int_0^{\pi}\cos^4\frac{1}{4}\theta\, d\theta$$

$$= 16a^2\int_0^{\pi/2}\cos^4\phi.\ 2\, d\phi,$$

$\left(\text{putting } \frac{1}{2}\theta = \phi \text{ so that } \frac{1}{2}d\theta = d\phi; \text{also when } \theta = 0,\right.$

$\left. \phi = 0 \text{ and when } \theta = \pi, \phi = \frac{1}{2}\pi \right)$

$$= 32a^2 \frac{3.1}{4.2} \cdot \frac{\pi}{2} = 6\pi a^2. \quad ...(1)$$

Area OACPO $= \frac{1}{2}\int_0^{\pi/2} r^2 \, d\theta$, for the parabola $r = \frac{2a}{1+\cos\theta}$

$$= \frac{1}{2} \cdot 4a^2 \int_0^{\pi/2} \frac{d\theta}{(1+\cos\theta)^2} = \frac{a^2}{2}\int_0^{\pi/2} \sec^4 \frac{1}{4}\theta \, d\theta$$

$$= \frac{1}{2}a^2 \int_0^{\pi/2} \left(1 + \tan^2 \frac{1}{2}\theta\right) \sec^2 \frac{1}{2}\theta \, d\theta$$

$$= a^2 \left[\tan\frac{1}{2}\theta + \frac{1}{3}\tan^3\frac{1}{2}\theta\right]_0^{\pi/2}$$

$$= a^2 \left(1 + \frac{1}{3}\right) = \frac{4a^2}{3}. \quad ...(2)$$

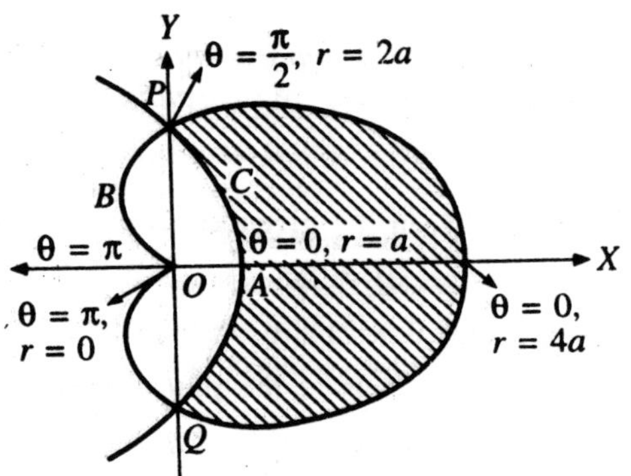

Fig 1.45

Also area OPBO $= \frac{1}{2}\int_{\pi/2}^{\pi} r^2 \, d\theta$, for the cardioid $r = 2a(1 + \cos\theta)$

$$= \frac{1}{2}\int_{\pi/2}^{\pi} 4a^2 (1 + \cos\theta)^2 \, d\theta$$

$$= 2a^2 \int_{\pi/2}^{\pi} [1 + 2\cos\theta + \cos^2\theta] \, d\theta$$

$$= 2a^2 \int_{\pi/2}^{\pi} \left(1 + 2\cos\theta + \frac{1}{2} + \frac{1}{2}\cos 2\theta\right) d\theta$$

$$= 2a^2 \left[\frac{3}{2}\theta + 2\sin\theta + \frac{1}{4}\sin 2\theta\right]_{\pi/2}^{\pi}$$

$$= \frac{3}{2}\pi\ a^2 - 4a^2. \qquad ...(3)$$

Adding (2) and (3) and multiplying by 2, we get the whole area included between the two curves *i.e.*, the area of the smaller portion of the cardioid

$$= 2 \times \left[\frac{4}{3}a^2 + \left(\frac{3}{2}\pi a^2 - 4a^2\right)\right] = a^2\left[3\pi - \frac{16}{3}\right]$$

$$= \frac{1}{3}a^2\ [9\pi - 16]. \qquad ...(4)$$

Also the shaded area (*i.e.*, the area of the larger portion of the cardioid)

= (Area of the whole cardioid) – (unshaded area)

i.e., = (1) – (4)

$$= 6\pi\ a^2 - \frac{1}{3}a^2\ (9\pi - 16) = \frac{1}{3}a^2\ (9\pi + 16). \qquad ...(5)$$

∴ Ratio of the two parts

$$= \frac{\text{Larger area}}{\text{Smaller area}} = \frac{\frac{1}{3}a^2(9\pi + 16)}{\frac{1}{3}a^2(9\pi - 16)} = \frac{9\pi = 16}{9\pi - 16}.$$

Example 68:

O is the pole of the lemniscate $r^2 = a^2 \cos 2\theta$ and PQ is a common tangent to its two loops. Find the area bounded by the line PQ and the area OP and OQ of the curve.

Solution:

The given curve is symmetrical about the initial line and also about the pole. The curve consists of two loops as shown in the figure.

From the figure it is clear that the common tangent to the two loops of the curve *i.e.*, the line PQ is parallel to the axis or x. Thus, the tangent PQ at the point P makes an angles π with the x-axis *i.e.*, $\psi = \pi$ at P.

Now differentiating the given equation $r^2 = a^2 \cos^2\theta$, we wet

$$2r\frac{dr}{d\theta} = a^2(-2\sin 2\theta), \text{ or } \frac{dr}{d\theta} = -\frac{a^2 \sin 2\theta}{r}$$

$$\therefore \quad \tan\phi = r\frac{d\theta}{dr} = -\frac{r \times r}{a^2 \sin 2\theta} = -\frac{a^2 \sin 2\theta}{a^2 \sin 2\theta}$$

$$= -\cot 2\theta = \tan\left(\frac{1}{2}\pi + 2\theta\right).$$

$$\therefore \quad \phi = \frac{1}{2}\pi + \theta^2.$$

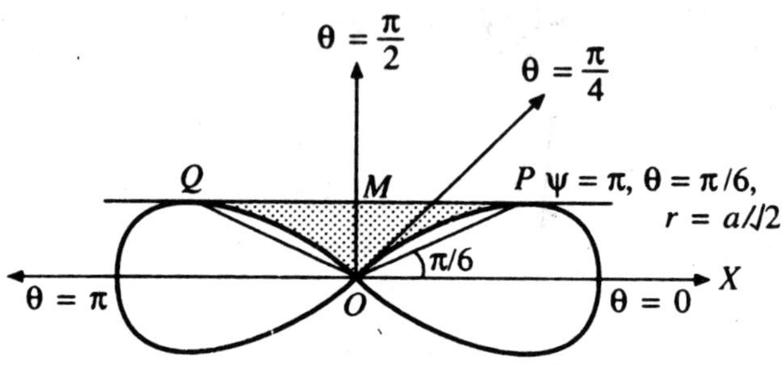

Fig 1.46

But for any point of the curve,

$$\psi = \theta + \phi + \left(\frac{1}{2}\pi + 2\theta\right) = \frac{1}{2}\pi + 3\theta.$$

Since at P, $\psi = \pi$, therefor at P, $\pi = \frac{1}{2}\pi + 3\phi$ or $\theta = \pi/6$.

Thus the vectorial angle θ of the point P is $\pi/6$ and the radius vector OP is given by $OP^2 = a^2 \cos\{2/6)\} = a^2/2$.

$$\therefore \quad OP = a\sqrt{2}.$$

Also putting $r = 0$ in the equation $r^2 = a^2 \cos 2\theta$, we get

$$\cos 2\theta = 0 \text{ or } 2\theta = \pm\pi/2 \text{ or } \theta = \pm\pi/4.$$

Thus $\theta = \pm\pi/4$ are the tangents at the pole to the curve.

Now the require area (*i.e.*, the dotted area)

= 2 [area of the ΔOPM – area of the segment OPSO of the curve $r^2 = a^2 \cos 2\theta$]

$$= 2\left[\frac{1}{2}OM.OP - \int_{\pi/}^{\pi/4} \frac{1}{2} r^2 d\theta, \text{ for the curve } r^2 = a^2 \cos 2\theta\right]$$

$$= 2\left[\frac{1}{2}\left(OP^2 \sin\frac{1}{6}\pi\right).\left(OP\cos\frac{1}{6}\pi\right) - \int_{\pi/6}^{\pi/4} \frac{1}{2} a^2 \cos 2\theta\right]$$

$$= 2\left[\frac{1}{2}OP^2 \sin\frac{1}{6}\pi \cos\frac{1}{6}\pi - \frac{a^2}{2}\left(\frac{\sin 2\theta}{2}\right)_{\pi/6}^{\pi/4}\right]$$

$$= 2\left[\frac{1}{2}\cdot\frac{a^2}{2}\cdot\frac{1}{2}\cdot\frac{\sqrt{3}}{2} - \frac{1}{4}a^2\left(1-\frac{\sqrt{3}}{2}\right)\right]$$

$$= 2a^2\left[\frac{\sqrt{3}}{16} - \frac{1}{4} + \frac{\sqrt{3}}{8}\right] = \frac{2a^2}{16}[\sqrt{3} - 4 + 2\sqrt{3}] = \frac{a^2}{8}(3\sqrt{3} - 4).$$

Example 69:

Find the area of a loop of the curve $r = a \sin 3\theta$ outside the circle $r = a/2$ and hence find the whole area of the curve outside the circle $r = a/2$.

Solution:

Eliminating r between the two given equations, we get $(a/2) = a \sin 3\theta$ *i.e.*, $\sin 3\theta = \frac{1}{2}$ *i.e.*, $3\theta = \pi/6$ or $5\pi/6$

i.e., $\theta = \pi/18$ or $5\pi/18$ *i.e.*, $\theta = 10°$ or $50°$

Thus the loop of the curve $r = a \sin 3\theta$ lying between $q = 0$ and $\theta = \pi/3$ intersects the circle $r = a/2$ at the points B and B' where $\theta = 10°$ at B and $\theta = 50°$ at B'. This loop is symmetrical about OA and $\theta = \pi/6$ at A.

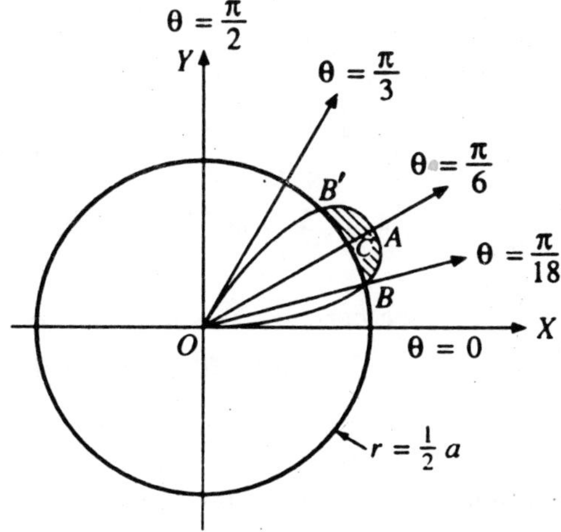

Fig 1.47

Now the required area of a loop of the curve $r = a \sin 3\theta$ lying outside the circle $r = a/2$ = the area BAB' CB

(*i.e.*, the shaded area)

= 2 × area BACB. (by symmetry)

= 2 × [(area of the curve $r = a \sin 3\theta$ between the radii vectors OB and OA *i.e.*, $\theta = \pi/18$ and $\theta = \pi/6$) – (area of the circle $r = a/2$ between the radii vectors OB and OC *i.e.*, $\theta = \theta/18$ and $\theta = \pi/6$)]

$$= 2\left[\frac{1}{2}\int_{\pi/18}^{\pi/6} r^2\,d\theta, \text{ for the curve } r = a\sin 3\theta - \frac{1}{2}\int_{\pi/18}^{\pi/6} r^2 d\theta, \text{ for the circle } r = \frac{a}{2}\right]$$

$$= \int_{\pi/18}^{\pi/6} a^2 \sin^2 3\theta\,d\theta - \int_{\pi/18}^{\pi/6} \frac{a^2}{4} d\theta$$

$$= \frac{a^2}{2}\int_{\pi/18}^{\pi/6} (1-\cos 6\theta) - \frac{a^2}{4}[\theta]_{\pi/18}^{\pi/6}$$

$$= \frac{a^2}{2}\left[\theta - \frac{\sin 6\theta}{6}\right]_{\pi/18}^{\pi/6} - \frac{a^2}{4}\left[\frac{\pi}{6} - \frac{\pi}{18}\right]$$

$$= \frac{a^2}{2}\left[\left\{\frac{\pi}{6} - \frac{\sin\pi}{6}\right\} - \left\{\frac{\pi}{18} - \frac{1}{6}\sin\frac{\pi}{3}\right\}\right] - \frac{a^2}{4}\cdot\frac{\pi}{9}$$

$$= \frac{a^2}{2}\left[\frac{\pi}{6} - \left\{\frac{\pi}{18} - \frac{1}{6}\cdot\frac{\sqrt{3}}{2}\right\}\right] - \frac{a^2\pi}{36}$$

$$= \frac{a^2}{2}\left[\frac{\pi}{9} + \frac{\sqrt{3}}{12}\right] - \frac{a^2\pi}{36} = \frac{a^2}{72}[2\pi + 3\sqrt{3}].$$

Again the curve $r = a \sin 3\theta$ has 3 equal loops

($\because$ n = 3 which is odd).

$\therefore$ whole area of the curve $r = a \sin 3\theta$ outside the circle $r = a/2$

= 3×area BAB' CB *i.e.*, 3 times the shaded area

$$= 3 \times \frac{1}{72}a^2[2\pi + 3\sqrt{3}] = \frac{1}{24}a^2[2\pi + 3\sqrt{3}].$$

Cartesian Equations Changed to Polar Form

Sometimes it is convenient to find the required area if the given cartesian equation of the curve is changed to polar form by putting $x = r \cos \theta$ and $y = r \sin \theta$.

Example 70:

Find the area of a loop of the folium

$$x^3 + y^3 = 3axy.$$

Also prove that the area included between the folium $x^3 + y^3 = 3axy$ and its asymptote is equal to the area of its loop.

Solution:

Changing the equation of the curve $x^3 + y^3 = 3axy$ into polar form by putting $x = r \cos \theta$ and $y = r \sin \theta$, we have

$$(r \cos \theta)^3 + (r \sin \theta)^3 = 3a\,(r \cos \theta)\times(r \sin \theta)$$

or $$r = 3a \cos \theta \sin \theta/(\cos^3\theta + \sin^3\theta). \qquad ...(1)$$

From (1), $r = 0$ when $\theta = 0$ and when $\theta = \pi/2$,

$\therefore$ the loop lies between $\theta = 0$ and $\theta = \pi/2$.

Hence the required area of the loop

$$= \frac{1}{2}\int_0^{\pi/2} r^2\, d\theta = \frac{1}{2}\int_0^{\pi/2}\left(\frac{3a\cos\theta\sin\theta}{\cos^3\theta+\sin^3\theta}\right)^2 d\theta$$

$$= \frac{9a^2}{2}\int_0^{\pi/2}\frac{\cos^2\theta\sin^2\theta}{(\cos^3\theta+\sin^3\theta)^2}\,d\theta$$

$$= \frac{9a^2}{2}\int_0^{\pi/2}\frac{\tan^2\theta\sec^2\theta}{(1+\tan^3\theta)^2}\,d\theta,$$

dividing the numerator and the denominator by $\cos^6\theta$.

Now put $1 + \tan^3\theta = t$

so that $3\tan^2\theta\sec^2\theta\, d\theta = dt$.

Also when $\theta = 0$, $t = 1$

and when $\theta \to \pi/2$, $t \to \infty$.

$$\therefore \text{ area of the loop } = \frac{9a^2}{2}\int_1^{\infty}\frac{1}{t^2}\cdot\frac{dt}{3} = \frac{3a^2}{2}\left[-\frac{1}{t}\right]_1^{\infty} = \frac{3a^2}{2}$$

Now to find asymptotes of the curve $x^3 + y^3 = 3axy$, putting $y = m$ and $x = 1$ in the highest degree terms, we get $\phi_3(m) = m^3 + 1$. The only real root of the equation $\phi_3(m) = 0$ *i.e.*, $m^3 + 1 = 0$ is, $m = -1$. Also $\phi_2(m) = -3am$. For $m = -1$, c is given by $c(3m^2) - 3am = 0$. Therefore, $c = -a$ when $m = -1$.

Hence $y = -x - a$ is the only real asymptote of the curve. Changing to polars, the equation of the asymptote $x + y + a = 0$ becomes $r = -a/(\cos\theta + \sin\theta)$. Considering its cartesian equation we find that the asymptote makes an angle $3\pi/4$ with the axis of x.

We shall now find the area between the curve and its asymptote. We shall first find the are between the curve and its asymptote lying in the second quadrant.

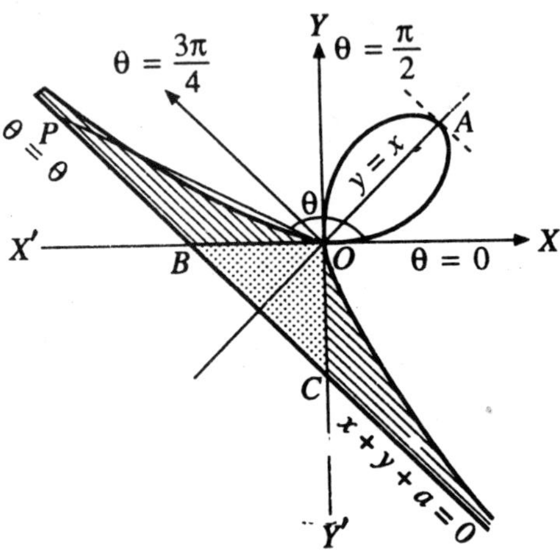

Fig 1.48

Let P be any point on the asymptote for which the vectorial angle θ lies between $3\pi/4$ and π. Join OP. As the point P tends to infinity along the asymptotes, $\theta \to 3\pi/4$. Now the area included between the folium (1) and its asymptote lying in the second quadrant

$= \lim_{\theta \to 3\pi/4}$ [{sectorial area between the asymptote and the radii vectors $\theta = \theta$ and $\theta = \pi$} – {sectorial area between the curve and the radii vectors $\theta = \theta$ and $\theta = \pi$}].

$$= \lim_{\theta \to 3\pi/4}\left[\left\{\frac{1}{2}\int_\theta^\pi r^2 d\theta, \text{ for the assymptote}\right\} - \left\{\frac{1}{2}\int_\theta^\pi r^2 d\theta, \text{ for the curve}\right\}\right]$$

$$= \lim_{\theta \to 3\pi/4}\left[\frac{a^2}{2}\int_\theta^\pi \frac{d\theta}{(\cos\theta + \sin\theta)^2} - \frac{9a^2}{2}\int_\theta^\pi \frac{\sin^2\theta\cos^2\theta\, d\theta}{(\cos^3\theta + \sin^3\theta)^2}\right]$$

$$= \lim_{\theta \to 3\pi/4}\left[\frac{a^2}{2}\int_\theta^\pi \frac{\sec^2\theta\, d\theta}{(1 + \tan\theta)^2}\right.$$ (dividing Nr. & Dr. by $\cos^2\theta$

$$\left. - \frac{3a^2}{2}\int_\theta^\pi \frac{3\tan^2\theta\sec^2\theta\, d\theta}{(1+\tan^2\theta)^2} \text{(dividing Nr. \& Dr. by } \cos^6\theta) \right]$$

$$= \lim_{\theta\to 3\pi/4}\left[\frac{a^2}{2}\left(-\frac{1}{1+\tan\theta}\right)_\theta^\pi - \frac{3a^2}{2}\left(-\frac{1}{1+\tan^3\theta}\right)_\theta^\pi\right]$$

$$= \lim_{\theta\to 3\pi/4}\left[\frac{a^2}{2}\left(-1+\frac{1}{1+\tan\theta}\right) - \frac{3a^2}{2}\left(-\frac{1}{1+\tan^3\theta}-1\right)\right]$$

$$= \lim_{\theta\to 3\pi/4}\left[\frac{a^2}{2}\left(\frac{-\tan\theta}{1+\tan\theta}\right) - \frac{3a^2}{2}\left(-\frac{\tan^3\theta}{1+\tan^3\theta}\right)\right]$$

$$= \lim_{\theta\to 3\pi/4}\left[\frac{a^2}{2}\cdot\frac{-\tan\theta(1-\tan\theta+\tan^2\theta)+3\tan^3\theta}{(1+\tan^3\theta)}\right]$$

$$= \lim_{\theta\to 3\pi/4}\left[\frac{a^2}{2}\cdot\frac{2\tan^3\theta+\tan^2\theta-\tan\theta}{1+\tan^3\theta}\right]$$

$$= \lim_{\theta\to 3\pi/4}\left[\frac{a^2}{2}\cdot\frac{(1+\tan\theta)(2\tan^2\theta-\tan\theta)}{(1+\tan\theta)(1-\tan\theta+\tan^2\theta)}\right]$$

$$= \lim_{\theta\to 3\pi/4}\left[\frac{a^2}{2}\cdot\frac{\tan\theta(2\tan\theta-1)}{1-\tan\theta+\tan^2\theta}\right] = \frac{a^2}{2}\cdot\left[\frac{3}{3}\right] = \frac{a^2}{2}.$$

Since the given curve is symmetrical about the line $y = x$, therefore, the area between the curve and the asymptote lying in the fourth quadrant is also equal to $a^2/2$. Also the area lying in the third quadrant, being that of a triangle, is $\frac{1}{2}$OB.OC *i.e.*, $\frac{1}{2}a^2$.

Hence the required area between the curve and its asymptote $= \frac{1}{2}a^2 + \frac{1}{2}a^2 + \frac{1}{2}a^2 = (3/2)a^2$ = the area of the loop of the curve.

Example 71:

Find the area of a loop of the curve $x^4 + y^4 = 4a^2xy$.

Solution:

Changing the equation of the curve $x^4 = y^4 = 4a^2xy$ into polar from by putting $x = r\cos\theta$,

$y = r\sin\theta$, we have

$r^4(\cos^4\theta + \sin^4\theta) = 4a^2r^2 \cos\theta \sin\theta$

or $r^2 = 4a^2 \sin\theta \cos\theta/(\cos^4\theta + \sin^4\theta)$. ...(1)

From (1), r = 0, when $\sin\theta = 0$

or $\cos\theta = 0$ *i.e.*, $\theta = 0$ or $\pi/2$.

∴ a loop lies between $\theta = 0$ and $\pi = \theta/2$.

Hence the required area of a loop

$$=\frac{1}{2}\int_0^{\pi/2} r^2 d\theta = \frac{1}{2}\int_0^{\pi/2} \frac{4a^2 \sin\theta\cos\theta}{\cos^4\theta + \sin^4\theta} d\theta, \text{ from (1)}$$

$$=2a^2\int_0^{\pi/2} \frac{\tan\theta\sec^2\theta}{1+\tan^4\theta} d\theta,$$

dividing the Nr. and Dr. by $\cos^4\theta$.

Now put $\tan^2\theta = t$

so that $2\tan\theta \sec^2\theta\, d\theta = dt$.

Also when $\theta = 0$, t = 0

and when $\theta \to \pi/2$, $t \to \infty$.

∴ the required area

$$= a^2\int_0^{\pi/2} \frac{dt}{1+t^2} = a^2[\tan^{-1}t]_0^{\infty}$$

$$= a^2[\tan^{-1}\infty - \tan^{-1}0] = a^2(\pi/2).$$

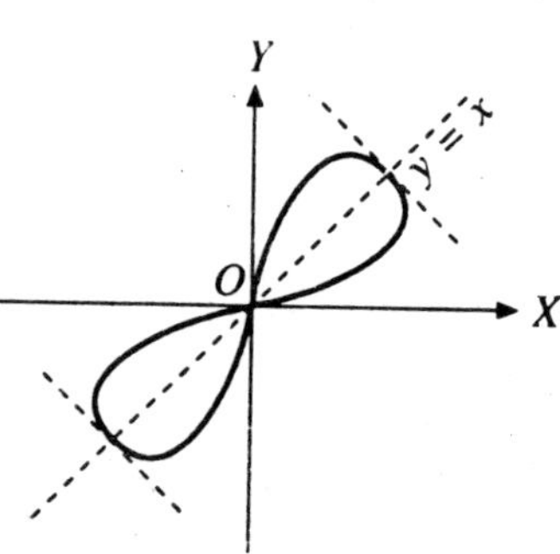

Fig 1.49

Example 72:

Find the area of a loop of the curve $(x^2 + y^2)^2 = 4axy^2$.

Solution:

Changing to polara by putting $x = r\cos\theta$, $y = r\sin\theta$, the equation of the curve becomes $(r^2)^2 = 4a(r\cos\theta)(r\sin\theta)^2$ o $r = 4a\cos q \sin^2 q$. ...(1)

From 91), r = 0 when $\sin^2\theta = 0$ or $\cos\theta = 0$ *i.e.*, $\theta = 0$ or $\pi/2$.

∴ a loop lies between $\theta = 0$ and $\theta = \pi^2$.

Hence the required area of loop

$$= \frac{1}{2}\int_0^{\pi/2} r^2 d\theta$$

$$= \frac{1}{2}\times 16a^2\int_0^{\pi/2} \cos^2\theta \sin^4\theta\, d\theta, \text{ from (1)}$$

$$= 8a^2 \times \frac{1.3.1}{6.4.2}\cdot\frac{\pi}{2} = \frac{\pi a^2}{4}.$$

Example 73:

Find the area of a loop the curve $(x^2 + y^2)^2 = 4xy^2$.

Solution:

Do your self.

Here a = 1.

Example 74:

Prove that the area of a loop of the curve

$$x^6 + y^6 = a^2x^2y^2 \text{ is } \pi\, a^2/12.$$

Solution:

Put x = r cos θ and y r sin θ to change the equation of the curve into polar form. Then the curve becomes

$$r^6 (\cos^6\theta + \sin^6\theta) = a^2r^4 \,.\, \cos^2\theta \sin^2\theta$$

or $$r^2 = a^2 \sin^2\theta\ \text{cso}^2\theta/(\cos^6\theta + \sin^6\theta). \quad ...(1)$$

From 91), r = 0 when $\sin^2\theta = 0$ or $\cos^2\theta = 0$ *i.e.*, θ = 0 or π/2.

∴ a loop lies between θ = 0 and θ = π/2.

hence the required area of the loop

$$= \frac{1}{2}\int_0^{\pi/2} r^2\, d\theta$$

$$= \frac{1}{2}\int_0^{\pi/2} \frac{a^2 \sin^2\theta \cos^2\theta d\theta}{\cos^6\theta + \sin^6\theta} \text{ from(1)}$$

$$= \frac{1}{2}\int_0^{\pi/2} \frac{\tan^2\theta \sec^2\theta}{1 + \tan^6\theta} d\theta, \text{ dividing the Nr. and Dr. by } \cos^6\theta.$$

Now put $\tan^3\theta = t$

so that $3 \tan^2\theta \sec^2\theta\, d\theta = dt$.

Also when θ = 0, t = 0

and when θ → π/2, t → ∞.

$$\therefore \text{ the required area } = \frac{a^2}{6}\int_0^{\infty} \frac{dt}{1+t^2} = \frac{a^2}{6}\left[\tan^{-1} t\right]_0^{\infty}$$

$$= \frac{1}{6}a^2\left[\tan^{-1}\infty - \tan^{-1} 0\right]$$

$$= \frac{1}{6}a^2 \cdot (\pi/2) = \frac{1}{12}\pi a^2.$$

Example 75:

Find the area of a loop of the curve $x^4 + 3x^2y^2 + 2y^4 = a^2xy$.

Solution:

Put x = r cos θ and y = r sin θ to change the equation of the curve into polar form. Then the curve becomes

$$r^4 (\cos^4\theta + 3\cos^2\theta \sin^2\theta + 2\sin^4\theta) = a^2r^2 \cos\theta \sin\theta$$

or
$$r^2 = \frac{a^2 \sin\theta \cos\theta}{\cos^4\theta + 3\cos^2\theta \sin^2\theta + 2\sin^4\theta}, \quad ...(1)$$

From (1)), r = 0 when sin θ = 0 or cos θ = 0, *i.e.*, θ = 0 or π/2.

∴ a loop lies between θ = 0 and θ = π/2

Hence the required area of the loop

$$= \frac{1}{2}\int_0^{\pi/2} r^2\, d\theta = \frac{1}{2}\int_0^{\pi/2} \frac{a^2 \sin\theta \cos\theta\, d\theta}{\cos^4\theta + 3\cos^2\theta \sin^2\theta + 2\sin^4\theta} \text{ from (1)}$$

$$= \frac{a^2}{2}\int_0^{\pi/2} \frac{\tan\theta \sec^2\theta d\theta}{1 + 3\tan^2\theta + 2\tan^4\theta}, \text{ dividing the Nr. and Dr. by } \cos 4\theta.$$

Now put $\tan^2\theta = t$

so that $2\tan\theta \sec^2\theta\, d\theta = dt$

and the new limits are t = 0 to t = ∞.

$$\therefore \text{ the required area } = \frac{a^2}{4}\int_0^{\infty} \frac{dt}{1 + 3t + 2t^2}$$

$$= \frac{a^2}{4}\int_0^{\infty} \frac{dt}{(1+t)9(1+2t)}$$

$$= \frac{a^2}{4}\int_0^{\infty} \left(\frac{2}{1+2t} - \frac{1}{1+t}\right)dt, \text{ by partial fractions}$$

$$= \frac{a^2}{4}[\log(1+2t) - \log(1+t)]_0^{\infty} = \frac{a^2}{4}\left[\log\left(\frac{1+2t}{1+t}\right)\right]_0^{\infty}$$

$$= \frac{a^2}{4} \lim_{t\to\infty} \log\left\{\frac{(1/t)+2}{(1/t)+1}\right\} - \frac{a^2}{4}\left[\log\left(\frac{1+2t}{1+t}\right)\right]_{t=0}$$

$$= \frac{1}{4}a^2[\log(2/1)] - \frac{1}{4}a^2[\log(1/1)] = \frac{1}{4}a^2 \log 2.$$

Example 76:

Prove that the area of a loop of the curve $x^5 + y^5 = 5ax^2y^2$ is five times the area of one loop of the curve $r^2 = a^2 \cos 2\theta$.

Solution:

Changing the equation of the curve $x^5 + y^5 = 5ax^2y^2$ into polar form by putting $x = r$ cso θ and $y = r \sin \theta$, we get

$$r^5 (\cos^5 \theta + \sin^5 \theta) = 5ar^4 \cos^2 \theta \sin^2 \theta$$

or $$r = a \sin^2 \theta \cos^2 \theta / (\cos^5 \theta + \sin^5 \theta). \quad ...(1).$$

From (1), $r = 0$

when $\sin^2 \theta = 0$

or $\cos^2 \theta = 0$ *i.e.*,

when $\theta = 0$ or $\theta = \pi/2$.

$\therefore$ The loop of the curve (1) lies between $\theta = 0$ and $\theta = \pi/2$.

Hence the required area of the loop of $x^5 + y^5 = 5ax^2y^2$

$$= \frac{1}{2}\int_0^{\pi/2} r^2 d\theta$$

$$\therefore \quad \frac{1}{2}\int_0^{\pi/2} \frac{25a^2 \sin^4 \theta \cos^4 \theta}{(\sin^5 \theta + \cos^5 \theta)^2} d\theta, \text{ from (1)}$$

$= \dfrac{25a^2}{2}\displaystyle\int_0^{\pi/2} \frac{\tan^4 \theta \sec^2 \theta d\theta}{(\tan^5 \theta + 1)^2}$, dividing the the Nr. and Dr. by $\cos^{10}\theta$.

Now put $\tan^5 \theta + 1 = t$

so that $5 \tan^4 \theta \sec^2 \theta \, d\theta = dt$.

Also when $\theta = 0$, $t = 1 + \tan 0 = 1$

and when $\theta \to \pi/2$, $t \to \infty$.

$$= \frac{25a^2}{2}\int_1^{\infty} \frac{1dt}{t^2 5} = \frac{5a^2}{2}\int_0^{\infty} \frac{1}{t^2} dt$$

$$= \frac{5a^2}{2}\left[\frac{-1}{t}\right]_1^{\infty} = \frac{5a^2}{2}.$$

The area of one loop of the curve

$$r^2 = a^2 \text{ cso } 2\theta = \frac{1}{2}a^2\cdot \text{ (Deduce it here).}$$

Thus the area of a loop of $x^5 + y^5 = 5ax^2y^2$

$$= \frac{5}{2}a^2 = 5 \cdot \left(\frac{1}{2}a^2\right) = 5\cdot$$

[area of a loop of $r^2 = a^2 \cos 2\theta$].

Example 77:

Prove that the area of the loop of the curve $x^3 + y^3 = 3axy$ is three times the area of loop of the curve $r^2 = a^2 \cos 2\theta$ (lemniscate).

Solution:

The area of a loop of $x^3 + y^3 = 3axy$ is $\frac{3}{2}a^2$. (Find it here)

And area of a loop of the lemniscate,

$$r^2 = a^2 \cos 2\theta, \text{ is } \frac{1}{2}a^2. \text{ (Find it here)}$$

Clearly area of the loop of $x^3 + y^3 = 3axy$

$$= \frac{3}{2}a^2 = 3\cdot\left(\frac{1}{2}a^2\right)$$

$$= 3 \text{ (area of a loop of } r^2 = a^2 \cos 2\pi).$$

Area of the Closed Curve

If a closed curve be given by the equations $x = f(t)$, $y = f(t)$, t being the parameter, then its area is given by

$$\frac{1}{2}\int_{t_1}^{t_2}\left(x\frac{dy}{dt} - y\frac{dx}{dt}\right)dt,$$

where t_1 and t_2 are the values of t such that the point travels completely once round the curve in anti-clockwise direction.

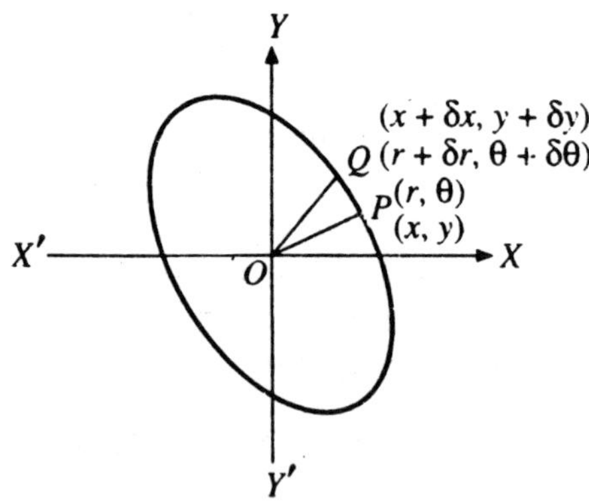

Fig 1.50

Suppose that the curve does not cut itself. Let P and Q be two neighbouring points on the curve. Let the cartesian Co-ordinates of P and

Q be (x, y) and (x + δx, y + δy) respectively and their polar Co-ordinates be (r, θ) and (r + δr, θ +δθ) respectively. Suppose that a point has to move anticlokwise in travelling from P to Q along the curve *i.e.*, the area falls on the left of an observer travelling along the curve from P to Q.

$$\text{Now area of } \Delta \text{ OPQ} = \frac{1}{2}\text{OP} \cdot \text{OP} \cdot \sin < \text{POQ}$$

$$= \frac{1}{2}r \times (r + \delta r) \sin \delta\theta$$

$$= \frac{1}{2}r^2 \delta\theta, \text{ neglecting higher powers of } \delta r \text{ and } d\theta.$$

$$\text{Again the area of triangle OPQ} = \frac{1}{2}\begin{vmatrix} 1 & 0 & 0 \\ 1 & x & y \\ 1 & x+\delta x & y+\delta y \end{vmatrix}$$

$$= \frac{1}{2}[x(y+\delta y) - y(x+\delta x)] = \frac{1}{2}[x\delta y - y\delta y].$$

$$\text{Thus the area of } \Delta \text{ OPQ} = \frac{1}{2}r^2\delta\theta = \frac{1}{2}(x\delta y - y\delta x).$$

$$\text{Hence the required area} = \frac{1}{2}\int r^2 d\theta \; \frac{1}{2}\int (xdy - ydx)$$

$$= \frac{1}{2}\int\left(x\frac{dy}{dt} - y\frac{dx}{dt}\right)dt,$$

the limits of integration are those values of t so that the point P (x, y) returns to its initial position.

Notes:

1. The area is positive or negative according as the direction of rotation is anticlockwise or clockwise when point P moves from P to Q *i.e.*, the area is +ive or –ive according as the area lies to the left or right of the point moving on the curve in the direction in which t increases.

2. Suppose the curve crosses itself so as the form a figure of eight as shown in the figure (i). Here two loops are described; one loop lies on the lift while the second lies on the right the direction in which t increases. Thus the above formula will given us the difference of the areas of the two loops. Hence to find the whole area in this case, we should find each area separately and then find their sum.

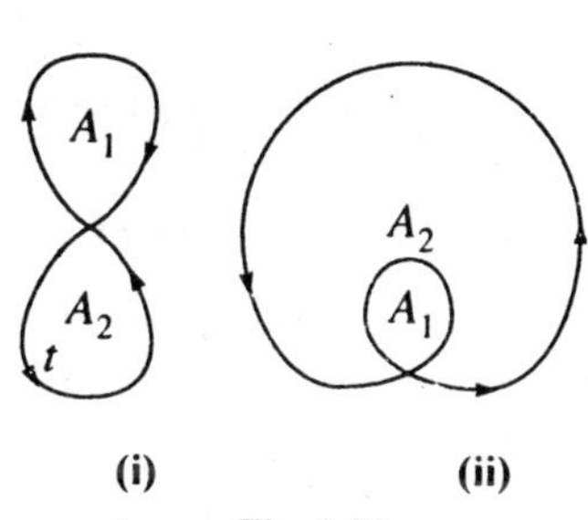

Fig 1.51

But if the curve cuts itself as in [fig. (ii)] then the area given by the above formula will be the sum of the area of the two loops *i.e.*, the area so obtained will be $A_1 + A_2$.

Example 78:

Find the area of the ellipse x = a cos t, y = b sin t.

Solution:

The ellipse is a closed curve and is completely described when t varies from 0 to 2π .

Also (dx/dt) = – a sin t and (dy/dt) = b cos t.

$$\therefore\ x\frac{dy}{dt} - y\frac{dx}{dt} = ab\ (\cos^2 t + \sin^2 t) = ab.$$

hence the required area

$$= \frac{1}{2}\int_0^{2\pi}\left(x\frac{dy}{dt} - y\frac{dx}{dt}\right)dt$$

$$= \frac{1}{2}\int_0^{2\pi} ab\ dt = \frac{1}{2}ab[t]_0^{2\pi}$$

$$= \frac{1}{2}ab\cdot 2\pi = \pi\ ab.$$

Example 79:

Show that area of the loop of the curve $x = a\ (1 - t^2),\ y = at\ (1 - t^2),\ -1 \le t \le 1$, *is* $8a^2/15$.

Solution:

Here x and y both vanish when t = – 1 and then again when t = 1. When – 1 < t < 1, x is always positive, but y is positive when t is +ive, and negative when t is –ive When t = 0, x = a and y = 0. Also by giving values to t equal in magnitude by t opposite in sign we observe that for each value of x between 0 and a there area two equal and opposite values of y.

Thus a loop (symmetrical about x-axis) is obtained as t varies from – 1 to 1.

∴ the required area of the loop

$$= \frac{1}{2}\int_{-1}^{1}\left(x\frac{dy}{dt} - y\frac{dx}{dt}\right)dt$$

$$= \frac{1}{2}\int_{-1}^{1}[a(1-t^2)(a-3at^2) - at(1-t^2)(-2at)]dt$$

$$= \frac{a^2}{2}\int_{-1}^{1}\left[1-3t^2-t^2+3t^4+2t^2-2t^4\right]dt$$

$$= \frac{a^2}{2}\int_{-1}^{1}(1-2t^2+t^4)dt = \frac{a^2}{2}\left(t-\frac{2t^2}{2}+\frac{t^5}{5}\right)_{-1}^{1}$$

$$= \frac{a^2}{2}\left(\frac{16}{15}\right) = \frac{8a^2}{15}$$

Example 80:

Find the area of the loop of the curve

$$x = \frac{a\sin 3\theta}{\sin\theta},\ y = \frac{a\sin 3\theta}{\cos\theta}.$$

Solution:

Here x and y both vanish when $\theta = -\pi/3$ and then again when $\theta = \pi/3$. When $-\pi/3 < \theta < 0$, x is +ive and y is –ive, when $0 < \theta < \pi/3$, x and y are both positive; and when $\theta \to 0$, $x \to 3a$ and $y \to 0$. Also we observe that for each value of x between 0 and 3a, there are two equal and opposite values of y . Thus, a loop is formed between $\theta = -\pi/3$ and $\theta = \pi/3$ and it is symmetrical about x -axis.

Note:

Obviously $\tan\theta = y/x$. Also

$$x = a\frac{3\sin\theta - 4\sin^3\theta}{\sin\theta} = a\,(3 - 4\sin^2\theta)$$

$$= a\left(3 - 4\frac{y^2}{x^2+y^2}\right)$$

or $$x\,(x^2+y^2) = a\,(3x^2 - y^2) \text{ or } y^2\,(a+x) = x^2\,(3a-x).$$

Thus, the cartesian equation of the above curve is $y^2(a+x) = x^2(3a - x)$ and x loop is formed between x = a and x = 3a, and x = – a is an asymptote]

∴ the required area of the loop

$$= \frac{1}{2}\int_{-\pi/3}^{\pi/3}\left(x\frac{dy}{d\theta} - y\frac{dx}{d\theta}\right)d\theta$$

$$= \frac{1}{2}\int_{-\pi/3}^{\pi/3}\left[\frac{a\sin 3\theta}{\sin\theta}\times a\left\{\frac{\cos\theta.(3\cos 3\theta)+\sin 3\theta\sin\theta}{\cos^2\theta}\right\}\right.$$

$$\left. - \frac{a\sin 3\theta}{cso\theta}\times a\left\{\frac{\sin\theta(3\cos 3\theta) - \sin 3\theta\cos\theta}{\sin^2\theta}\right\}\right]d\theta$$

$$= \frac{a^2}{4}\int_{-\pi/3}^{\pi/3}\frac{\sin^2 3\theta}{(\cos^2\theta\sin^2\theta)}d\theta = \frac{a^2}{4}\int_{-\pi/3}^{\pi/3}\frac{(3\sin\theta - 4\sin^3\theta)^2}{\sin^2\theta\cos^2\theta}d\theta$$

$$= \frac{a^2}{4}\int_{-\pi/3}^{\pi/3}\frac{(3-4\sin^2\theta)^2}{\cos^2\theta}d\theta = \frac{a^2}{4}\int_{-\pi/3}^{\pi/3}\frac{\{3-4(1-\text{cso}^2\theta)\}^2}{\cos^2\theta}d\theta$$

$$= \frac{a^2}{4}\cdot\int_{-\pi/3}^{\pi/3}\frac{(4\cos^2\theta - 1)^2}{\cos^2\theta}d\theta = \frac{a^2}{4}\int_{-\pi/3}^{\pi/3}(16\text{cso}^2\theta - 8 + \sec^2\theta)\,d\theta$$

$$= \frac{a^2}{4}\cdot 2\int_0^{\pi/3}\left[8(1+\cos 2\theta) - 8 + \sec^2\theta\right]d\theta$$

$$= \frac{a^2}{4}\int_0^{\pi/3}\left[8\cos 2\theta + \sec^2\theta\right]d\theta$$

$$= \frac{a^2}{4}\left[4\sin 2\theta + \tan\theta\right]_0^{\pi/3}$$

$$= \frac{a^2}{4}\left[4\cdot\frac{\sqrt{3}}{2} + \sqrt{3}\right] = \frac{3\sqrt{3a^2}}{2}.$$

Example 81:

Prove that area of the curve $x = a\cos\theta = b\sin\theta + c$, $y = a'\,\text{cso}\,\theta + b'\sin\theta + c'$ si equal to $\pi\,(ab' - a'b)$.

Solution:

To find the shape of the curve shift the origin to the point (c, c'). Then the equations of the curve become $x = a\cos\theta + b\sin\theta$, $y = a'\cos\theta + b'\sin\theta$.

These equations are linear in cos and sin θ, and if solved, the values of cos θ, sin θ will involve x an y in the first degree and will be homogenous in x and y. Hence on squaring and adding, the equation to the curve will be of the form $Ax^2 + 2Hxy + By^2 = 1$ which is a conic with its center at the new origin.

From the given equations of the curve we observe that x and y are both finite for all values of θ. Thus, the given conic is a closed curve and it is completely traced from $\theta = 0$ to $\theta = 2\pi$.

$$\therefore \text{ the required area } = \frac{1}{2}\int_0^{2\pi}\left(x\frac{dy}{d\theta} - y\frac{dx}{d\theta}\right)d\theta$$

$$= \frac{1}{2}\int_0^{2\pi}[(a\cos\theta + b\sin\theta + c)(-a\sin\theta + b'\cos\theta)$$

$$- (a'\cos\theta + b\sin\theta + c')(-a\sin\theta + b\cos\theta)]$$

$$= \frac{1}{2}\int_0^{2\pi} [ab'(\cos^2\theta + \sin^2\theta) - a'b(\sin^2\theta + \cos^2\theta)$$
$$+ c(b'\cos\theta - a'\sin\theta) + c'(\sin\theta - b\cos\theta)]d\theta$$
$$= \frac{1}{2}\int_0^{2\pi} (ab' - a'b)d\theta + \frac{1}{2}(cb' - c'b)\int_0^{2\pi} \cos\theta d\theta$$
$$= \frac{1}{2}(c'a - ca')\int_0^{2\pi} \sin\theta d\theta$$
$$= \frac{1}{2}(ab' - a'b)[\theta]_0^{2\pi} + \frac{1}{2}(cb' - c'b)[\sin\theta]_0^{2\pi}$$
$$+ \frac{1}{2}(c'a - ca')[-\cos\theta]_0^{2\pi}$$
$$= \pi\ (ab' - a'b).$$

Example 82:

If P (x, y) be any point on the ellipse $x^2/a^2 + y^2b^2 = 1$ and S be the sectorial area bounded by the curve, the x-axis and the line joining the origin to P, show that x = a cos (2S/ab), y = b sin (2S/ab).

Solution:

The parametric equations of the ellipse are

$$x = a \cos t,\ y = b \sin t. \qquad ...(1)$$

At the vertex A, t = 0 and at P, t = t.

$$S = \text{sectorial area AOP} = \frac{1}{2}\int_{t=0}^{t}\left(x\frac{dy}{dt} - y\frac{dx}{dt}\right)dt$$
$$= \frac{1}{2}\int_{t=0}^{t}[a\cos t \cdot b\cos t - b\sin t \cdot (-a\sin t)]dt$$
$$= \frac{1}{2}ab\int_0^t(\cos^2 t + \sin^2 t)dt$$
$$= \frac{1}{2}ab\int_0^t dt = \frac{1}{2}ab[t]_0^t = \frac{1}{2}abt.$$

$\therefore$ t = (2S/ab). Putting this value of t in (1), we get

x = a cos (2S/ab) and y = b sin (2S/ab).

Example 83:

If A is the vertex, O is the centre and P any point on the hyperbola $x^2/a^2 - y^2/b^2 = 1$, show that x = a cosh (2S/ab) and y = b sinh (2S/ab), where S is the sectorial area OPA.

Solution:

The parametric equations of the hyperbola are x = a cosh t, y = b sinh t ...(1)

At the vertex A, t = 0 and at P, t = t.

$$\therefore\ S = \text{the sectorial area AOP}$$

$$= \frac{1}{2}\int_0^1\left(x\frac{dy}{dt} - y\frac{dx}{dt}\right)dt$$

$$= \frac{1}{2}\int_0^1(a\cosh t\cdot b\text{coht} - b\sinh t\cdot a\sinh t)\,dt$$

$$= \frac{1}{2}ab\int_0^t(\cosh^2 t - \sinh^2 t)dt$$

$$= \frac{1}{2}ab\int_0^t dt = \frac{1}{2}ab\int_0^t \frac{1}{2}\,adt$$

$\therefore$ t = (2S/ab). Putting this value of t in (1),

we get x = a cosh (2S/ab), y = b sinh (2S/ab).

Example 84:

If the pedal equation of the a curve be p f(r), prove that the area bounded by the curve and two radii vectors is $\frac{1}{2}\int\frac{pr\,dr}{\sqrt{(r^2-p^2)}}$, *taken between suitable limits.*

Solution:

The required area = $\frac{1}{2}\int r^2 d\theta$, between the suitable limits.

$$\text{Now } \tan\phi = \frac{rd\theta}{dr}.$$

$$\text{Also } \sin\phi = \frac{p}{r}$$

$$\Rightarrow \quad \tan\phi = \frac{p}{\sqrt{(r^2-p^2)}}.$$

$$\therefore \quad \frac{r\,d\theta}{dr} = \frac{p}{\sqrt{(r^2-p^2)}}$$

i.e.,

$$r^2\,d\theta = \frac{pr\,dr}{\sqrt{(r^2-p^2)}}.$$

$$\therefore \quad \text{required area} = \frac{1}{2}\int r^2 d\theta$$

$$= \frac{1}{2}\int \frac{pr\,dr}{\sqrt{(r^2 - p^2)}},$$ between the suitable limits.

Example 85:

Find the area bounded by the axis of x, and the following curves and the given and the given ordinates:

(i) $y = c \cosh (x/c)$; $x = 0$, $x = a$.

(ii) $y = \sin^2 x$ $x = 0$, $x = 1/2\ \pi$.

(iii) $y = \log x$; $x = a$, $x = b$ $(b > a > 1)$.

(iv) $xy = c^2$; $x = a$, $x = b$, $(a > b > 0)$.

Solution:

(i) The required area $= \int_0^a y\,dx$,

($\therefore$ limits are form $x = 0$ to $x = a$)

$$= \int_0^a c \cosh\left(\frac{x}{c}\right)dx = c^2\left[\sinh\frac{x}{c}\right]_0^a = c^2 \sin h\ \frac{a}{c}.$$

(ii) The required area $= \int_0^{\pi/2} y\,dx = \int_0^{\pi/2} \sin^2 x\,dx$,

($\because$ $y = \sin^2 x$)

$$= \frac{1}{2}\int_0^{r/2} (1 - \cos 2x)\,dx = \frac{1}{2}\left[x - \frac{\sin 2x}{2}\right]_0^{\pi/2}$$

$$= \frac{1}{2}\left[\frac{\pi}{2} - \frac{1}{2}\times 0\right] = \frac{\pi}{4}.$$

(iii) The required area $= \int_a^b y\,dx = \int_a^b \log x\,dx$

$$= \left[(\log x).x\right]_a^b - \int_a^b \frac{1}{x}.\,x\,dx, \text{ integrating by parts}$$

$= [b \log b - a \log a] - [b - a]$

$= (b \log b - b) - (a \log a - a)$

$= b \log (b/e) - a \log (a/e)$, $[\because \log e = 1]$

(iv) The required area $\int_b^a y dx = \int_b^a (c^2/x)\,dx$, $[\therefore\ y = c^2/x]$

$$= c^2.\left[\log x\right]_b^a = c^2 (\log a - \log b) = c^2 \log (a/b).$$

Example 86:

Find the area bounded by the curve $y = x^3$, the y-axis and the lines $y = 1$ and $y = 8$.

Solution:

Here the curve is bounded between the axis of y and the lines parallel to x-axis. Hence the required area

$$= \int_{y=1}^{8} x\,dy$$

$$= \int_{1}^{8} y^{1/3}\,dy, \qquad [\because x^3 = y \text{ or } x = y^{1/3}]$$

$$= \left[\frac{3}{4}y^{4/3}\right]_1^8 = \frac{3}{4}[8^{4/3} - 1^{4/3}] = \frac{3}{4}[2^4 - 1] = \frac{45}{4}.$$

Example 87:

Find the area bounded by the ellipse

$$\frac{x^2}{a^2} + \frac{y^2}{b^2} = 1$$

The ordinates $x = c$, $x = d$ and the x-axis.

Solution:

Equation of the ellipse is

$$\frac{x^2}{a^2} + \frac{y^2}{b^2} = 1$$

or $$\frac{y^2}{b^2} = 1 - \frac{x^2}{a^2}$$

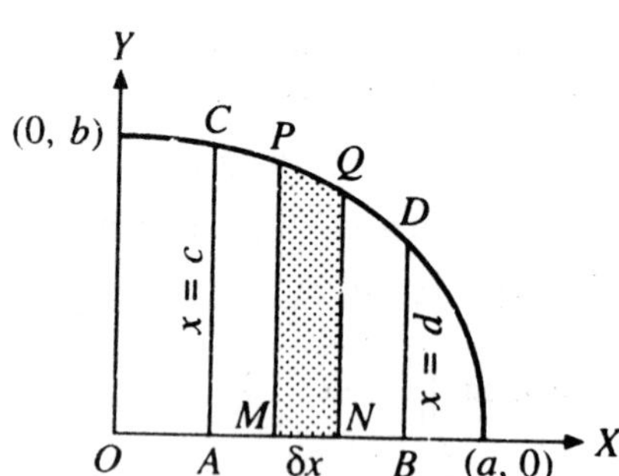

Fig 1.52

given $y = \frac{b}{a} \sqrt{(a^2 - x^2)}$ ...(1)

$\therefore$ the required area = the area

$$\text{ABDC} = \int_c^d y \, dx$$

$$= \int_c^d \frac{b}{a} \sqrt{(a^2 - x^2)} \, dx, \quad \text{from (1)}$$

$$= \frac{b}{a}\left[\frac{1}{2} x \sqrt{(a^2 - x^2)} + \frac{1}{2} a^2 \sin^{-1}\left(\frac{x}{a}\right)\right]_c^d.$$

$$= \frac{b}{2a}\left[d\sqrt{(a^2 - d^2)} - c\sqrt{(a^2 - c^2)} + a^2\left(\sin^{-1}\frac{d}{a} - \sin^{-1}\frac{c}{a}\right)\right]$$

Example 88:

Find the area of the quadrant of an ellipse.

Solution:

Here the required area (*i.e.*, the area of a quadrant) lies between the limits $x = 0$ and $x = a$.

$\therefore$ area of the quadrant $= \int_0^a y \, dx$

$$= \frac{b}{a}\int_0^a \sqrt{(a^2 - x^2)} \, dx, \quad \text{from (1) or Ex. 3 (a)}$$

$$= \frac{b}{a}\left[\frac{1}{2} x \sqrt{(a^2 - x^2)} + \frac{1}{2} a^2 \sin^{-1}\left(\frac{x}{a}\right)\right]_0^a$$

$$= \frac{b}{a}\left[\frac{1}{2}(0 - 0) + \frac{1}{2} a^2\left(\frac{1}{2}\pi - 0\right)\right] = \frac{\pi ab}{4}.$$

Example 89:

Find the area bounded by the parabola $y^2 = 4ax$ and its latus rectum.

Solution:

Latus rectum is a line through the focus S (a, 0) and perpendicular to x-axis *i.e.*, its equation is $x = a$. Also the curve is symmetrical about x-axis.

$\therefore$ the required area LOL'

$$= 2 \times \text{area OSL} = 2.\int_0^a y \, dx$$

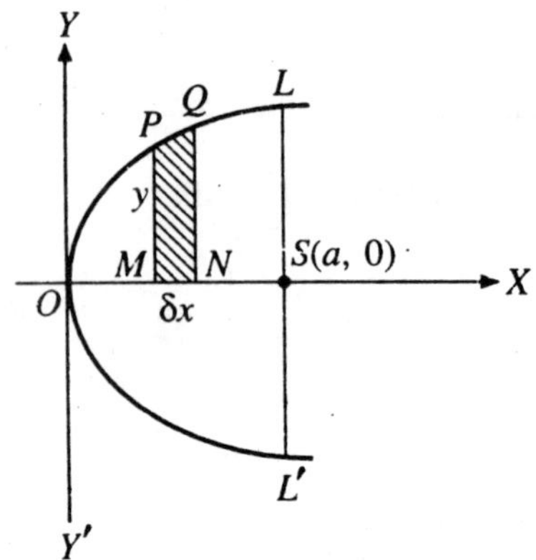

Fig 1.53

$$= 2\int_0^a \sqrt{(4ax)}\, dx, \quad [\because y^2 = 4ax, \textit{i.e.}, y = \sqrt{(4ax)}]$$

$$= 2\sqrt{(4a)}\left[\frac{2}{3}x^{3/2}\right]_0^a$$

$$= \frac{8}{3}\sqrt{(a)}.\ a^{3/2} = \frac{8}{3}a^2.$$

Example 90:

Find the whole area of the ellipse $(x^2/a^2) + (y^2/b^2) = 1$.

Solution:

Clearly, the area of an ellipse is 4 times the area of a quadrant.

$\therefore$ The required area of the ellipse

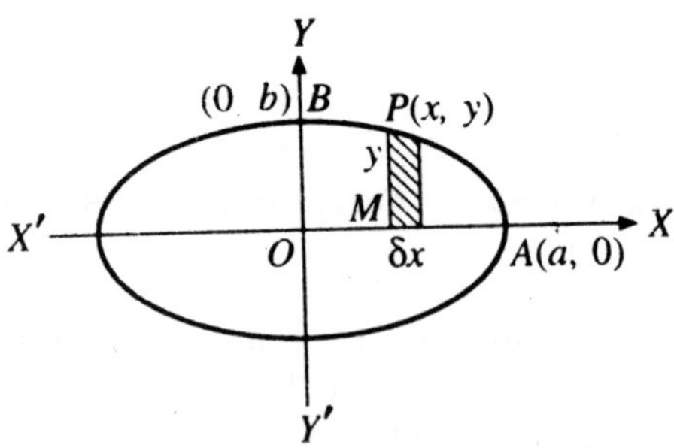

Fig 1.54

$$= 4 \int_0^a y\,dx$$

$$= 4.\frac{\pi ab}{4},$$

$$= \pi\ ab.$$

Example 91:

Show that the area cut off a parabola by any double ordinate is two third of the corresponding rectangle contained by that double ordinate and its distance from the vertex.

Solution:

Let the parabola be $y^2 = 4ax$. Also let $x = b$ be any double ordinate. Since the curve is symmetrical about x-axis, therefore the area cut off the parabola $y^2 = 4ax$ by the double ordinate $x = b$ is $2 \times$ (area included between the x-axis, $x = b$ and curve in the +ive quadrant).

$\therefore$ the required are $= 2 \int_{x=0}^{b} y\,dx$

$$= 2\int_0^b \sqrt{(4ax)}\,dx, \qquad [\because\ y^2 = 4ax]$$

$$= 4\sqrt{(a)}\left[\frac{2}{3}x^{3/2}\right]_0^b = \frac{8}{3}\sqrt{(a)}.\,b^{3/2} \qquad ....(1)$$

Again, at $x = b$, from $y^2 = 4ax$, we have

$$y^2 = 4ab \text{ or } y = 2\sqrt{(ab)}.$$

$\therefore$ length of the double ordinate $= 2y = 2.2\sqrt{(ab)} = 4\sqrt{(ab)}$.

Now area of the rectangle contained by the double ordinate and its distance from the vertex $= 2y \times x = 2y\,.b$

$$= 4\sqrt{(ab)}.\,b = 4a^{1/2}\,b^{3/2}.$$

Its two third $= \frac{2}{3}.\ 4a^{1/2}\,b^{3/2}$

$$= \frac{8}{3}a^{1/2}\,b^{3/2}$$

= area cut off a parabola by the double ordinate, from (1).

Example 92:

Trace the curve $ay^2 = x^2(a - x)$ and show that the area of its loop is $8a^2/15$.

Solution:

Tracing:

(i) The curve is symmetrical about x-axis and passes through the origin.

(ii) Equating to zero the lowest degree terms in the equation of the curve, we get $y^2 - x^2 = 0$ *i.e.*, $y = \pm x$ as the tangents at origin and these being real and distinct the node is expected at the origin.

(iii) At $y = 0$, we $x = 0$ and $x = a$ *i.e.*, the curve crosses the x-axis at (0, 0) and (a, 0). Also when $x > a$, y^2 is negative, *i.e.*, y is imaginary. Hence the curve does not exist for values of $x > a$.

Also as x decreases from 0 to $-\infty$, y increases from 0 to ∞.

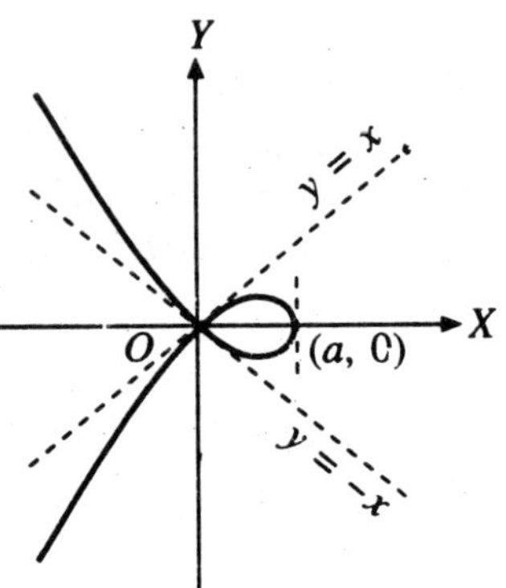

Fig 1.55

(iv) Now asymptotes. Thus, the shape of the curve is as shown in the figure. Clearly the loop is formed between $x = 0$ and $x = a$.

$\therefore$ required area of the loop $= 2\int_0^a y\,dx$,

[$\because$ curve is symmetrical about x-axis]

$= 2\int_0^a \frac{x\sqrt{(a-x)}}{\sqrt{}}dx$, putting for y from the equation of the curve

$= 2\int_0^{\pi/2} \frac{a\sin^2\theta\sqrt{(a)}.\cos\theta}{\sqrt{a}} 2a\sin\theta\cos\theta\,d\theta$,

[putting $x = a\sin^2\theta$, $dx = 2a\sin\theta\cos\theta\,d\theta$]

$= 4a^2\int_0^{\pi/2} \sin^3\theta\cos^2\theta\,d\theta$

$$= 4a^2 \frac{\Gamma 2.\Gamma\frac{3}{2}}{2\Gamma\frac{7}{2}} = 4a^2.\frac{\frac{1}{2}\sqrt{\pi}}{2.\frac{5}{2}.\frac{3}{2}.\frac{1}{2}\sqrt{\pi}} = \frac{8}{15}a^2.$$

Example 93:

Find the area of the loop of the curve $3ay^2 = x(x-a)^2$.

Solution:

The curve is symmetrical about x-axis. Putting $y = 0$, we get $x = 0$ and $x = a$, *i.e.*, the loop is formed between $x = 0$ and $x = a$.

∴ required area of the loop

$$= 2\int_0^a y\,dx, \qquad [\because \text{ curve is symmetrical about x-axis}].$$

Now for the portion of the loop lying in the first quadrant, y is +ive and x lies between 0 and a. Therefore for this portion of the loop, we have $y = \{1/\sqrt{(3a)}\}.\ \sqrt{(x)}.\ (a - x)$.

∴ required area of the loop

$$= 2\int_0^a \frac{(a-x).\sqrt{x}}{\sqrt{(3a)}},$$

putting for y from the given equation of the curve

$$= \frac{2}{\sqrt{(3a)}}\int_0^a (ax^{1/2} - x^{3/2})\,dx$$

$$= \frac{2}{\sqrt{(3a)}}\left[\frac{2}{3}ax3/2 - \frac{2}{5}x5/2\right]_0^a = 8a^2/(15/\sqrt{3}).$$

Example 94:

Find the area of the loop of the curve $y^2 = x(x-1)^2$.

Solution:

Here also the curve is symmetrical about x-axis. Putting y = 0, we get x = 0 and x = 1 *i.e.*, the loop is formed between x = 0 and x = 1.

∴ required area of the loop $= 2\int_0^1 y\,dx$

$= 2\int_0^1 (1-x).\ \sqrt{x}\,dx$, putting for y from the equation of the curve.

[Note that we have taken, $y = \sqrt{(x)}.\ (1 - x)$]

$$= 2\int_0^1 (x^{1/2} - x^{3/2})\,dx = 2\left[\frac{2}{3}x^{3/2} - \frac{2}{5}x^{5/2}\right]_0^1$$

$$= 2\left[\frac{2}{3} - \frac{2}{5}\right] = \frac{8}{15}.$$

Example 95:

Find the area of a loop of the curve $xy^2 + (x + a)^2 (x + 2a) = 0$.

Solution:

The curve is symmetricals $x^2 (x + a) = 0$

or $$y^2 = \frac{x^2 (a+x)}{a-x} \quad ...(1)$$

Note that the shifting of the origin only changes the equation of the curve and has no effect on its shape. Now the origin being at the points A, the new limits for the loop are x = – a to x = 0.

∴ required area of the loop = 2 × area CPA

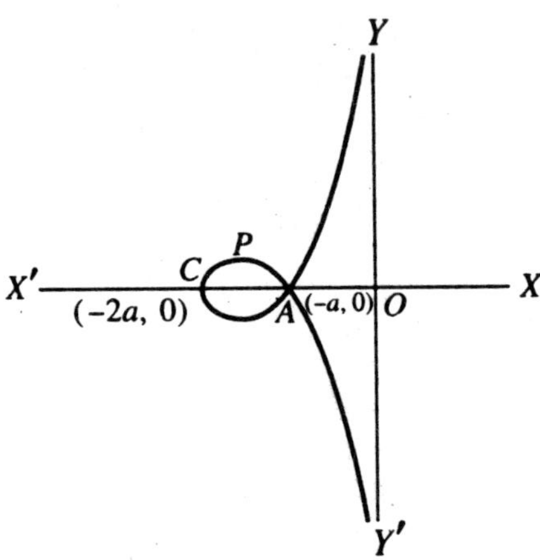

Fig 1.56

$$= 2 \int_{-a}^{0} y\, dx, \text{ [the value of y to be put from (1)]}$$

$$= 2 \int_{-a}^{0} \left\{ -x \sqrt{\left(\frac{a+x}{a-x} \right)} \right\} dx,$$

[Note that in the equation (1), for the portion

CPA, $y = -x\sqrt{\{(a + x)/(a - x)\}}$]

$$= 2 \int_{-a}^{0} \frac{-x(a+x)}{\sqrt{(a^2 - x^2)}} dx, \text{ multiplying the numerator}$$

and the denominator by $\sqrt{(a + x)}$

$$= 2 \int_{\pi/2}^{0} \frac{-(-a \sin\theta)(a - a\sin\theta)}{a\cos\theta} . (- a \cos \theta)\, d\theta,$$

$$= - 2a^2 \int_{\pi/2}^{0} (\sin \theta - \sin^2 \theta)\, d\theta$$

$$= 2a^2 \int_{0}^{\pi/2} (\sin \theta - \sin^2 \theta)\, d\theta$$

$$= 2a^2 \left[1 - \frac{1}{2}.\frac{1}{2}\pi\right], \qquad \text{(by Walli's formula)}$$

$$= 2a^2 \left(1 - \frac{1}{4}\pi\right).$$

Example 96:

Find the area between the curve y^2 (4 – x) = x^2 and its asymptote.

Solution:

The curve is symmetrical about the x-axis. It cuts the x-axis at x = 0 *i.e.*, at the origin. The straight line x = 4 is the asymptote of the curve.

$\therefore$ required area $= 2\int_0^4 y\,dx = 2\int_0^4 \sqrt{\left(\frac{x^2}{4-x}\right)}\,dx$, putting for y form the equation of the curve

$$= 2\int_0^4 \frac{x}{\sqrt{(4-x)}}dx$$

$$= 2\int_0^{\pi/2} \frac{4\sin^2\theta}{\sqrt{(4-4\sin^2\theta)}}.\ 8\sin\theta\cos\theta\,d\theta,$$

putting $x = 4\sin^2\theta$

so that $dx = 8\sin\theta\cos\theta\,d\theta$

$$= 32\int_0^{\pi/2} \sin3\theta\,d\theta = 32.\frac{2}{3.1}, \text{ (by Walli's formula)}$$

$$= \frac{64}{3} \text{ units of area.}$$

Example 97:

Find the area :

(i) of the loop of the curve

$$x(x^2 + y^2) = a(x^2 - y^2) \text{ or } y^2(a + x) = x^2(a - x).$$

(ii) of the portion bounded by the curve and its asymptotes.

Solution:

The curve is symmetrical about x-axis. The tangents at origin are $a(y^2 - x^2) = 0$ *i.e.*, $y = \pm x$. Since there are two real and distinct tangents at the origin, therefore the origin is a node on the curve.

Putting y = 0,

we get x = 0

and x = a *i.e.*, the loop is formed between x = 0 and x=a.

∴ required area of the loop

$=2\int_0^a y\,dx$, by symmetry

$=2\int_0^a x\sqrt{\left(\dfrac{a-x}{a+x}\right)}\,dx$, putting for y from the given equation of the curve

$=2\int_0^a \dfrac{x(a-x)}{\sqrt{(a^2-x^2)}}dx$, multiplying the numerator and the denominator by $\sqrt{(a-x)}$

$=2\int_0^{\pi/2} \dfrac{a\sin\theta(a-\sin\theta)}{\sqrt{(a^2-a^2\sin^2\theta)}}.\,a\cos\theta\,d\theta$,

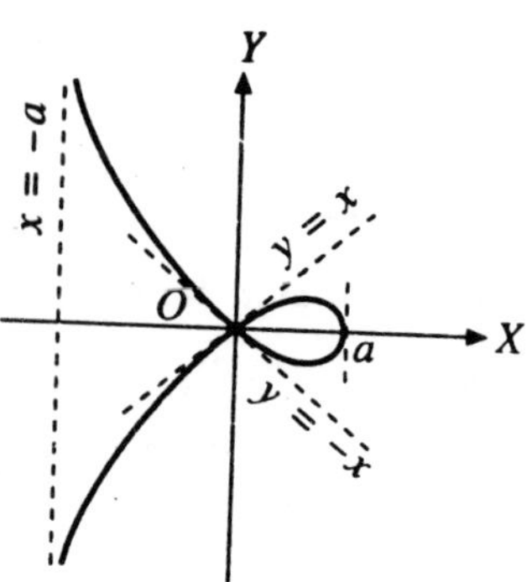

Fig 1.57

putting $x = a\sin\theta$ so that $dx = a\cos\theta\,d\theta$

$$= 2a^2\int_0^{\pi/2}\sin\theta\,(1-\sin\theta)\,d\theta$$

$$= 2a^2\left[\int_0^{\pi/2}\sin\theta d\theta - \int_0^{\pi/2}\sin^2\theta d\theta\right]$$

$$= 2a^2\left[1-\frac{1}{2}.\frac{1}{2}\pi\right], \text{ by Walli's formula}$$

$$= 2a^2\left(1-\frac{1}{4}\pi\right) = \frac{1}{2}a^2\,(4-\pi).$$

(ii) The line $x = -a$ is the asymptote of the curve.

Now the area lying between the curve and its asymptote

$=2\int_{-a}^0 y dx$, the value of y to be put from the equation of the curve

$$= 2\int_{-a}^0 -x\sqrt{\left(\frac{a-x}{a+x}\right)}\,dx,$$

[Note that for the arc of the curve lying in the second quadrant x is –ve and y is +ive so that $y = -x\sqrt{\{(a-x)/(a+x)\}}$ for this arc].

$$= 2\int_{-a}^0 \frac{-x(a-x)}{\sqrt{(a^2-x^2)}}\,dx$$

$$= 2\int_{\pi/2}^0 \frac{-(-a\sin\theta)(a+\sin\theta)}{\sqrt{(a^2-a^2\sin^2\theta)}}.\,(-a\cos\theta)\,d\theta,$$

putting x = – a sin θ

so that dx = – a cos θ dθ

$$= -2a^2 \int_{\pi/2}^{0} \sin\theta\,(1 + \sin\theta)\,d\theta$$

$$= 2a^2 \int_{0}^{\pi/2} (\sin\theta + \sin^2\theta)\,d\theta$$

$$= 2a^2\left[1 + \frac{1}{2}.\frac{1}{2}\pi\right], \text{ by Walli's formula.}$$

$$= 2a^2\left[1 + \frac{1}{4}\pi\right] = \frac{1}{2}a^2\,(4 + \pi).$$

Example 98:

Trace the curve $y^2\,(2a - x) = x^3$ and find the entre area between the curve and its asymptotes.

Solution:

Tracing of the curve $y^2\,(2a - x) = x^3$.

(i) Since in the equation of the curve the powers of y that occur are all even, therefore the curve is symmetrical about the axis of x.

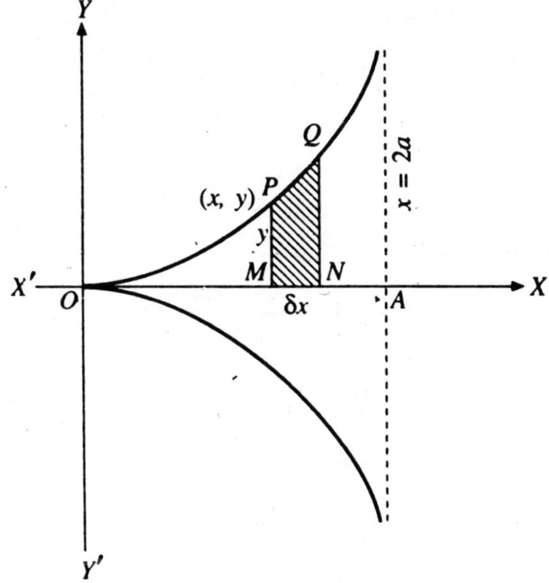

Fig 1.58

(ii) The curve passes through the origin. Equating to zero the lowest degree terms in the equation of the curve, we get the tangents at the origin as $2ay^2 = 0$ *i.e.*, $y = 0$. $y = 0$ are two coincident tangents at the origin. Therefore the origin may be a cusp.

(iii) The curve cuts the coordinate axes only at the origin.

(iv) Solving the equation of the curve for y, we get

$$y^2 = \frac{x^3}{2a - x}.$$

When $x = 0$, $y^2 = 0$.

When $x \to 2a$, $y^2 \to \infty$.

Therefore $x = 2a$ is an asymptote of the curve.

When $0 < x < 2a$, y^2 is +ive *i.e.*, y is real.

Therefore the curve exists in this region.

When $x > 2a$, y^2 is –ive *i.e.*, y is imaginary. Therefore, the curve does not exist in the region $x > 2a$.

When $x < 0$, y^2 is –ive. Therefore the curve does not exist in the region $x < 0$.

Combining all these facts, we see that the shape of the curve is as shown in the figure.

Now the required area = 2 × area in the first quadrant

$$= 2\int_0^{2a} y\,dx$$

$$= 2\int_0^{2a} \frac{x^{3/2}}{\sqrt{(2a - x)}}\,dx, \qquad \left[\because y^2 = \frac{x^3}{2a - x}\right]$$

Now put $x = 2a\sin^2\theta$, so that $dx = 4a\sin\theta\cos\theta\,d\theta$.

∴ the required area

$$= 2\int_0^{\pi/2} \frac{(2a\sin^2\theta)^{3/2}}{\sqrt{(2a - 2a\sin^2\theta)}}\cdot 4a\sin\theta\cos\theta\,d\theta$$

$$= 16a^2\int_0^{\pi/2} \frac{\sin^3\theta}{\cos\theta}\sin\theta\cos\theta\,d\theta$$

$$= 16a^2\int_0^{\pi/2} \sin 4\theta\,d\theta$$

$$= 16a^2.\frac{3.1}{4.2}.\frac{\pi}{2}, \qquad \text{(by Walli's formula)}$$

$$= 3\pi a^2.$$

2

Lengths of Curves

(Rectification)

2.1 Definition

The process of finding the length of an arc of a curve between two given points is called rectification.

2.2 Lengths of Curves

If s denotes the arc length of a curve measured from a fixed point to any point on it, we have

$$\frac{ds}{dx} = \pm\sqrt{\left\{1+\left(\frac{dy}{dx}\right)^2\right\}},$$

where +ive or –ive sing is to be taken before the radical sign according as x increases or decreases as s increases. Hence, if s increases as x increases, we have

$$= \frac{ds}{dx} = \sqrt{\left\{1+\left(\frac{dy}{dx}\right)^2\right\}}$$

or $$ds = \sqrt{\left\{1+\left(\frac{dy}{dx}\right)^2\right\}}\,dx.$$

Integrating, we have

$$s = \int_a^x \sqrt{\left\{1+\left(\frac{dy}{dx}\right)^2\right\}}\,dx,$$

where a is the abscissa of the fixed point from which s is measured.

Hence, the arc length of the curve $\mathbf{y} = \mathbf{f(x)}$ included between two points for which x = a and x = b is equal to

$$\int_a^b \sqrt{\left\{1+\left(\frac{\mathbf{dy}}{\mathbf{dx}}\right)^2\right\}}\,\mathbf{dx},\ (\mathbf{b} > \mathbf{a}).$$

Sometimes, it is more convenient to take y as the independent variable. Then the length of the arc of the cu rve **x = f (y)** between y = a and y = b is equal to

$$\int_a^b \sqrt{\left\{1+\left(\frac{dy}{dx}\right)^2\right\}}\,dy, \ (b > a).$$

Remark:

Suppose we have to find the length of the arc of a curve (whose cartesian equation is given) lying between the points (x_1, y_1) and (x_2, y_2). We can use either of two formulae

$$s = \int_{x_1}^{x_2} \sqrt{\{1+(dy/dx)^2\}}dx \text{ and}$$

$$s = \int_{y_1}^{y_2} \sqrt{\{1+(dx/dy)^2\}}dy.$$

If we feel any difficulty in integration while using one of these two formulae, we must try the other formula also.

Equations of the Curve in Parametric Form

If the equations of the curve be given in the parametric form x = f (t), y = φ (t), then s is obviously a function of t. In this case if we measure the arc length s in the direction of t increasing, we gave

$$\frac{ds}{dt} = \sqrt{\left\{\left(\frac{dx}{dt}\right)^2+\left(\frac{dy}{dt}\right)^2\right\}} \text{ or}$$

$$ds = \sqrt{\left\{\left(\frac{dx}{dt}\right)^2+\left(\frac{dy}{dt}\right)^2\right\}}\,dt.$$

On integrating between proper limits, the required length

$$s = \int_{t_1}^{t_1} \sqrt{\left\{\left(\frac{dx}{dt}\right)^2+\left(\frac{dy}{dt}\right)^2\right\}}\,dt.$$

Equation of the Curve in Polar Form

For the curve r = f (θ), if we measure the arc length s in the direction of θ increasing,, we have

$$\frac{ds}{d\theta} = \sqrt{\left\{r^2+\left(\frac{dr}{d\theta}\right)^2\right\}}$$

or $$ds = \sqrt{\left\{r^2 + \left(\frac{dr}{d\theta}\right)^2\right\}} d\theta.$$

On integrating between proper limits, the required length

$$s = \int_{\theta_1}^{\theta_2} \sqrt{\left\{r^2 + \left(r\frac{dr}{dq}\right)^2\right\}} d\theta$$

If the equation of the curve be $\theta = f(r)$, then the required length is given by

$$s = \int_{r_1}^{r_2} \sqrt{\left\{1 + \left(r\frac{d\theta}{dr}\right)^2\right\}} dr$$

Equation of the Curve in Pedal Form

Let $p = f(r)$ be the equation of the curve and r_1 and r_2 be the values of r at two given points of the curve. Then by differential calculus we know that

$$= \frac{ds}{dr} = \frac{r}{\sqrt{(r^2 - p^2)}}$$

or $$ds = \frac{r}{\sqrt{(r^2 - p^2)}} dr,$$

where s increases as r increases.

On integrating between proper limits. The required length

$$s = \int_{r_1}^{r_2} \frac{\mathbf{r}}{\sqrt{(\mathbf{r}^2 - \mathbf{p}^2)}} \mathbf{dr}.$$

The value of p should be put in terms of r from the equation of the curve.

Remark:

If the curve is symmetrical about one or more lines, then find out the length of one symmetrical part and then multiply it by the number of symmetrical parts.

2.3 Intrinsic Equations

Definition : By the intrinsic equation of a curve we mean a relation between s and ψ, where s is the length of the arc AP of the curve measured from a fixed point A on it to a variable point P, and ψ is the angle which the tangent to the curve at P makes with a fixed straight line usually taken as the positive direction of the axis of x. The co-ordinates s and ψ are known as *intrinsic Co-ordinates.*

(a) To Find the Intrinsic Equations from the Cartesian Equation

Let the equation of the given curve be $y = f(x)$. Take A as the fixed point on the curve from which s is measured and take the axis of x as the fixed straight line with reference to which ψ is measured. Let P (x, ψ) be any point on the curve and PT be the target at the point P to the curve.

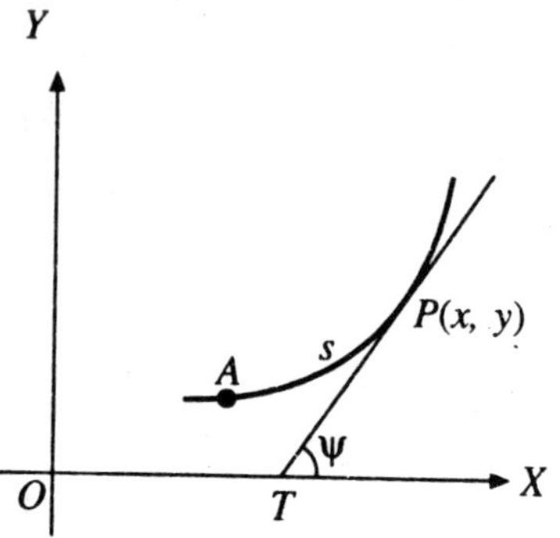

Fig 2.1

Let arc AP = s

and $\angle$ PTX = ψ.

Now, we have tan

$\psi = dy/dx = f'(x)$....(1)

Let a be the abscissa of the point A from which s is measured.

Then $$s = \int_a^x \sqrt{\left[1 + \left(\frac{dy}{dx}\right)^2\right]}dx$$

$$= \int_a^x \sqrt{\left[1 + \{f'(x)\}^2\right]}dx. \qquad ...(2)$$

Eliminating x between (1) and (2),

we obtain the required intrinsic equation.

Note: To find the intrinsic equation from the parametric equations we use $\frac{dy}{dx} = \frac{dy/dt}{dx/dt}$ and then proceed as in case (a).

(b) Intrinsic Equation from Polar Equation

Let the equation of the given curve be $r = f(\theta)$. Take A as the fixed point on the curve from which s is measured. Let P be any point (r, θ) on the curve.

Let arc AP = s and $\angle$ PTX = ψ, where OX is the initial line.

If ϕ is the angle between the radius vector and the tangent at P, then

$$\tan\phi = r\frac{d\theta}{dr} = \frac{r}{dr/d\theta} = \frac{f(\theta)}{f'(\theta)}, \qquad ...(1)$$

and $$y = \theta + \phi. \qquad ...(2)$$

Let α be the vectorial angle of the point A. Then we have

$$s = \int_{\alpha}^{\theta} \sqrt{\left\{r^2 + \left(\frac{dr}{d\theta}\right)^2\right\}} d\theta$$

$$= \int_{\alpha}^{\theta} \sqrt{\left[\{f(\theta)\}^2 + \{f'(\theta)\}^2\right]} d\theta \qquad ...(3)$$

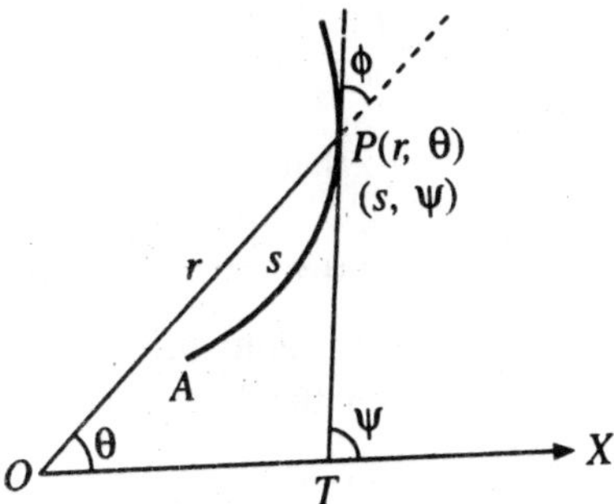

Fig 2.2

Eliminating θ and ϕ between (1), (2) and (3), we get a relation between s and ψ, which is the intrinsic equation of the curve.

(c) Intrinsic Equation from Pedal Equation

Let the pedal equation of the curve be $p = f(r)$. ...(1)

Then $$s = \int_a^r \frac{r\,dr}{\sqrt{(r^2 - p^2)}}, \qquad ...(2)$$

the arc length s being measured from the point r = a.

Also the radius of curvature $\rho = \dfrac{ds}{d\psi} = r\dfrac{dr}{dp}$.

Eliminating p and r between (1), (2) and (3), we obtain the required intrinsic equation.

MISCELLANEOUS EXAMPLES

Example 1:

Find the intrinsic equation of the cardioid $r = a(1 - \cos\theta)$.

Solution:

The given curve is $r = a(1 - \cos\theta)$.

Differentiating (1) w.r.t. θ, we have $dr/d\theta = a \sin\theta$.

$$\therefore \quad \tan\phi = r\frac{d\theta}{dr} = \frac{r}{dr/d\theta}$$

$$= \frac{r}{a\sin\theta} = \frac{a(1-\cos\theta)}{a\sin\theta}$$

$$= \frac{2\sin^2\frac{1}{2}\theta}{2\sin\frac{1}{2}\theta\cos\frac{1}{2}\theta} = \tan\frac{1}{2}\theta.$$

Therefore $\phi = \frac{1}{2}\theta$, so that

$$y = \theta + \phi = \theta + \frac{1}{2}\theta = \frac{3}{2}\theta, \text{ giving } \theta = \frac{2}{3}\psi. \quad ...(2)$$

If s denotes the arc length of the cardioid measured from the cusp O (*i.e.*, the point θ = 0) to any point P (r, θ) in the direction of θ increasing, we have

$$s = \int_0^\theta \sqrt{\left\{r^2 + \left(\frac{dr}{d\theta}\right)^2\right\}}d\theta = a\int_0^\theta \sqrt{\{(1-\cos\theta)^2 + \sin^2\theta\}}d\theta$$

$$= 2a\int_0^\theta \sin\frac{1}{2}\theta d\theta = 4a\left[-\cos\frac{1}{2}\right]_0^\theta$$

$$= 4a\left(1-\cos\frac{1}{2}\theta\right) = 8a\sin^2\frac{1}{6}\psi$$, which is the required intrinsic equation.

Example 2:

Find the length of the loop of the curve $x = t^2$, $y = t - 1/3t^3$.

Solution:

Eliminating the parameter t from $x = t^2$ and $y = t - 1/3t^3$, we get $y^2 = x\left(1-\frac{1}{3}x\right)^2$ as the cartesian equation of the curve and hence we observe that the curve is symmetrical about the x-axis. The loop of the curve extends from the point (0, 0) to the point (3, 0). Putting y = 0 in $y = t - \frac{1}{3}t^3$, we get t = 0 and $t = \frac{1}{2}\pi^3$. Therefore the arc of the upper half of the loop extends from t = 0 to $t = \sqrt{3}$.

Now the required length of the loop

= 2 × length of the half of the loop which lies above x-axis

$$= 2\int_0^{\sqrt{3}} \sqrt{\left\{\left(\frac{dx}{dt}\right)^2 + \left(\frac{dy}{dt}\right)^2\right\}}dt$$

$$= 2\int_0^{\sqrt{3}} \sqrt{\left\{(2t)^2 + \left(1 - \frac{1}{3}\cdot 3t^2\right)^2\right\}}dt$$

$$= 2\int_0^{\sqrt{3}} \sqrt{(1 + 2t^2 + t^4)}dt = 2\int_0^{\sqrt{3}} (1 + t^2)dt$$

$$= 2\left[t + \frac{t^3}{3}\right]_0^{\sqrt{3}} = 2[\sqrt{3} + \sqrt{3}] = 4\sqrt{3}.$$

Example 3:

Find the length of the arc of the curve $x = e^t \sin t$, $y = e^t \cos t$, *from* $t = 0$ *to* $t = \frac{1}{2}\pi$.

Solution:

The given equation of the curve are $x = e^t \sin t$, $y = e^t \cos t$.

Differentiating w.r.t. t, we have $dx/dt = e^t \cos t + e^t \sin t = e^t (\cos t + \sin t)$

and $\qquad dy/dt = e^t (-\sin t) + e^t \cos t = e^t (\cos t - \sin t)$.

$\therefore$ the required arc length $= \int_0^{\pi/2} \sqrt{\left\{\left(\frac{dx}{dt}\right)^2 + \left(\frac{dy}{dx}\right)^2\right\}}dt$

$$= \int_0^{\pi/2} \sqrt{\{e^{2t}(\cos t + \sin t)^2 + e^{2t}(\cos t - \sin t)^2\}}dt$$

$$= \int_0^{\pi/2} e^2 \sqrt{\{2(\cos^2 t + \sin^2 t)\}}dt = \sqrt{2}\int_0^{\pi/2} e^t dt = \sqrt{2}\left[e^t\right]_0^{\pi/2}$$

$$= \sqrt{2}\,[e^{\pi/2} - 1].$$

Example 4:

In the ellipse $x = a \cos \phi$, $y = b \sin \phi$, *show that* $ds = a\sqrt{(1 - e^2 \cos^2 \phi)}\, d\phi$, *and hence show that the whole length of the ellipse is*

$$2\pi a\left[1 - \left(\frac{1}{2}\right)^2 \cdot \frac{e^2}{1} - \left(\frac{1.3}{2.4}\right)^2 \cdot \frac{e^4}{3} - \left(\frac{1.3.5}{2.4.6}\right)^2 \frac{e^6}{5} - \ldots\right],$$ *where e is the eccentricity of the ellipse.*

Solution:

The given equation of the ellipse are $x = a \cos \phi$, $y = b \sin \phi$.

We have $dx/d\phi = -a \sin\phi$,

$dy/d\phi = b \cos\phi$.

If we measure the length s in the direction of ϕ increasing

$$\frac{ds}{d\phi} = \sqrt{\left\{\left(\frac{dx}{d\phi}\right)^2 + \left(\frac{dy}{d\phi}\right)^2\right\}}$$

$$= \sqrt{(a^2 \sin^2\phi + b^2 \cos^2\phi)}$$

$$= \sqrt{\{a^2 \sin^2\phi + a^2(1 - e^2)\cos^2\phi\}},$$

[$\because$ for ellipse, $b^2 = a^2(1 - e^2)$]

$$= a\sqrt{(1 - e^2\cos^2\phi)}.$$

$$\therefore \quad ds = a\sqrt{(1 - e^2\cos^2\phi)}\,d\phi. \quad ...(1)$$

Also the ellipse is symmetrical about both the axes and in the first quadrant ϕ varies from 0 to $1/2\pi$. Therefore whole length of the ellipse

$= 4 \times$ length of the ellipse lying in the first quadrant

$$= 4\int_0^{\pi/2} a\sqrt{(1 - e^2\cos^2\phi)}\,d\phi, \text{ [from (1)]}$$

$$= 4a\int_0^{\pi/2} (1 - e^2\cos^2\phi)a^{1/2}\,d\phi$$

$$= 4a\int_0^{\pi/2}\left[1 - \frac{1}{2}e^2\cos^2\phi - \frac{1}{2.4}e^4\cos^4\phi - \frac{1.3}{2.4.6}e^6\cos^6\phi - ...\right]d\phi$$

(on expanding by binomial theorem)

$$= 4a\left[\int_0^{\pi/2} 1\cdot d\phi - \frac{e^2}{2}\int_0^{\pi/2}\cos^2\phi\,d\phi - \frac{e^4}{2.4}\int_0^{\pi/2}\cos^4\phi\,d\phi - \frac{1.3}{2.4.6}e^6\int_0^{\pi/2}\cos^6\phi\,d\phi - ...\right]$$

$$= 4a\left[\frac{\pi}{2} - \frac{e^2}{2}\cdot\frac{1}{2}\cdot\frac{\pi}{2}\cdot - \frac{e^4}{2.4}\,\frac{3.1}{4.2}\cdot\frac{\pi}{2} - \frac{1.3}{2.4.6}e^6\cdot\frac{5.3.1}{6.4.2}\cdot\frac{\pi}{2} - ...\right]$$

$$= 2a\pi\left[1 - \frac{1}{2^2}e^2 - \frac{1^2\cdot 3}{2^2\cdot 4^2}e^4 - \frac{1^2\cdot 3^2\cdot 5}{2^2\cdot 4^2\cdot 6^2}e^6 - ...\right]$$

$$= 2\pi a\left[1 - \left(\frac{1}{2}\right)^2\cdot\frac{e^2}{1} - \left(\frac{1\cdot 3}{2\cdot 4}\right)^2\cdot\frac{e^4}{3} - \left(\frac{1\cdot 3\cdot 5}{2\cdot 4\cdot 6}\right)^2\cdot\frac{e^6}{5} - ...\right]$$

Example 5:

Show that in the epi-cycloid for which x = (a+b) cos θ–b cos{(a+b)/b}θ, y = (a + b) sin θ – b sin {(a + b)/b} θ, the length of the arc measured from the point θ = πb/a is {4b (a + b)/a} cos {(a/2b)θ }.

Solution:

Differentiating the given equation of the epicycloid w.r.t. θ,

we have dx/dθ = (a + b) (–sin θ) – b [–sin {(a + b)/b}θ] × [(a + b)/b]

= –(a + b) [sinθ –sin {(a + b)/b}θ] and dy/dθ

= (a + b) (cos θ) –b [cos {(a + b)/b}θ] × {(a + b)/b}

= (a + b) [cos θ – cos {(a + b)/b}θ].

We have $\left(\frac{ds}{d\theta}\right)^2 = \left(\frac{dx}{d\theta}\right)^2 + \left(\frac{dy}{d\theta}\right)^2$

$$= (a+b)^2\left[\left\{\sin\theta - \sin\frac{a+b}{b}\theta\right\}^2 + \left\{\cos\theta - \cos\frac{a+b}{b}\theta\right\}^2\right]$$

$$= (a+b)^2\left[\sin^2\theta + \sin^2\frac{a+b}{b}\theta - 2\sin\theta\sin\frac{a+b}{b}\theta\right.$$

$$\left. + \cos^2\theta + \cos^2\frac{a+b}{b}\theta - 2\cos\theta\cos\frac{a+b}{b}\theta\right]$$

$$= (a+b)^2\left[1 + 1 - 2\left(\sin\theta\sin\frac{a+b}{b}\theta + \cos\theta\cos\frac{a+b}{b}\theta\right)\right]$$

$$= 2(a+b)^2\left[1 - \cos\frac{a}{b}\theta\right]$$

$$= 4(a+b)^2\sin^2\frac{a}{2b}\theta. \qquad ...(1)$$

If s denodes the arc length of the epicycloid measured from the point θ = πb/a to the point θ = θ in the direction of θ decreasing, then s increases as θ decreases. Therefore ds/dθ will be negative. So taking square root of both sides of (1) and keeping the negative sign, we have

$$\frac{ds}{d\theta} = -2(a+b)\sin\frac{a}{2b}\theta$$

or $$ds = -2(a+b)\sin(a/2b)\theta\, d\pi.$$

The required length s is now given by

$$s = -\int_{b\pi/a}^{\theta} 2(a+b)\sin(a/2b)\theta d\theta$$

$$= -2(a+b)\,\frac{2b}{a}\left[-\cos\frac{a\theta}{2b}\,\theta\right]_{b\pi/a}^{\theta}$$

$$= \frac{4b(a+b)}{a}\left[\cos\frac{a\theta}{2b} - \cos\frac{\pi}{2}\right]$$

$$= \frac{4b(a+b)}{a}\cos\frac{a}{2b}\theta.$$

Example 6:

An ellipse of small eccentricity has its perimeter equal to that of a circle of radius r. Show that its area is $\pi r^2\left(1-\frac{3}{32}e^4\right)$ *nearly.*

Solution:

The parametric equations of the ellipse are $x = a\cos\phi$, $y = b\sin\phi$.

We have the perimeter of the ellipse

$$= 2a\pi\left[1-\left(\frac{1}{2}\right)^2 e^2 - \left(\frac{1\cdot 3}{2\cdot 4}\right)^2\cdot\frac{e^4}{3} - \ldots\right]$$

$$= 2a\pi\left[1-\frac{1}{4}e^2 - \frac{3}{64}e^4 - \ldots\right] = 2\pi r \quad \text{(as given)}$$

Therefore $$a = r\left[1-\left(\frac{1}{4}e^2+\frac{3}{64}e^4\right)\right]^{-1}, \qquad \ldots(1)$$

retaining upto e^4 because e is small.

Now area of the ellipse

$$= \pi ab = \pi a\times a\sqrt{(1-e^2)}, \qquad [\because b^2 = a^2(1-e^2)]$$
$$= \pi a^2(1-e^2)1/2$$

$$= \pi r^2\left[1-\left(\frac{1}{4}e^2+\frac{3}{64}e^4\right)\right]^{-2}\cdot(1-e^2)^{1/2} \quad \text{form (1)}$$

$$= \pi r^2\left[1+2\cdot\left(\frac{1}{4}e^2+\frac{3}{64}e^4\right)+3\left(\frac{1}{4}e^2+\frac{3}{64}e^4\right)^2+\ldots\right]$$

$$\times\left(1-\frac{1}{2}e^2-\frac{1}{8}e^4-\ldots\right), \text{ expanding by binomial theorem}$$

$$= \pi r^2\left(1+\frac{1}{2}e^2+\frac{3}{32}e^4+\frac{3}{16}e^4\right)\left(1-\frac{1}{2}e^2-\frac{1}{8}e^4\right),$$

retaining terms only upto e^4

$$= \pi r^2 \left(1 + \frac{1}{2}e^2 + \frac{9}{32}e^4\right)\left(1 - \frac{1}{2}e^2 - \frac{1}{8}e^4\right)$$

$$= \pi r^2 \left(1 + \frac{1}{2}e^2 + \frac{1}{8}e^4 + \frac{1}{2}e^2 - \frac{1}{4}e^4 + \frac{9}{32}e^4\right)$$

$$= \pi r^2 \left(1 - \frac{3}{32}e^4\right) \text{nearly.}$$

Example 7:

Find the length of the curve $x = e^t \left(\sin\frac{1}{2}t + 2\cos\frac{1}{2}t\right)$, $y = e^t \left(\frac{1}{2}t - 2\sin\frac{1}{2}t\right)$ *measured from* $t = 0$ *to* $t = \pi$.

Solution:

Differentiating the given parametric equations w.r.t. t,

we get $dx/dt = e^t \left(\cos\frac{1}{2}\cos\frac{1}{2}t - \sin\frac{1}{2}t\right) + e^t \left(\sin\frac{1}{2}t + 2\cos\frac{1}{2}t\right)$

$$= \frac{5}{2}e^t \cos\frac{1}{2}t$$

and $dy/dt = e^t \left(-\frac{1}{2}\sin\frac{1}{2}t - \cos\frac{1}{2}t\right) + e^t \left(\cos\frac{1}{2}t - 2\sin\frac{1}{2}t\right)$

$$= -\frac{5}{2}e^t \sin\frac{1}{2}t.$$

$\therefore$ the required length $= \int_0^\pi \sqrt{\left\{\left(\frac{dx}{dt}\right)^2 + \left(\frac{dy}{dt}\right)^2\right\}}dt$

$$= \int_0^\pi \sqrt{\left\{\left(\frac{5}{2}e^t \cos\frac{1}{2}t\right)^2 + \left(-\frac{5}{2}e^t \sin\frac{1}{2}t\right)^2\right\}}dt$$

$$= \frac{5}{2}\int_0^\pi e^t \sqrt{\left(\cos^2\frac{1}{2}t + \sin^2\frac{1}{2}t\right)}dt$$

$$= \frac{5}{2}\int_0^\pi e^t dt = \frac{5}{2}\left[e^t\right]_0^\pi$$

$$= \frac{5}{2}[e^\pi - 1].$$

Example 8:

Show that the length of the arc measured from t = 0 to any point t, of the curve x = a (cos t + t sin t), y = a (sin t – t cos t) is 1/2at³.

Solution:

Here $dx/dt = a\,[-\sin t + (1\times \sin t + t\cos t)] = a\,(t\cos t)$

and $dy/dt = a\,[\cos t - (1\times \cos t - t\times \sin t)] = a\,(t\sin t)$.

$\therefore$ the required length $= \int_0^t \sqrt{\left\{\left(\frac{dx}{dt}\right)^2 + \left(\frac{dy}{dt}\right)^2\right\}}dt$

$$= \int_0^t at\,\sqrt{\left(\sin^2 t + co^2 t\right)}dt = \int_0^t at\,dt = \frac{1}{2}at^2.$$

Example 9:

Prove that the arc of the curve given by x = a ×(3 sin t – sin³ t), y = a cos³ t measured from (0, a) to any point (x, y) is 3/2a (t + sin t cos t).

Solution:

Do your self.

Here t = 0 at the point (0, a) and t = 1 at the point (x, y).

Example 10:

Show that the length of an arc of the curve x sin t + y cos t = f '(t), x cos t – y sin t = f "(t) is given by s = f (t) + f "(t), where c is the constant of integration.

Solution:

The given equation of the curve are

$$x\sin t + y\cos t = f'(t) \quad ...(1)$$

and
$$x\cos t - y\sin t = f''(t). \quad ...(2)$$

Multiplying (1) by sin t and (2) by cost and adding, we get $x(\sin^2 t + \cos^2 t) = \sin t.\ f'(t) + \cos t.\ f''(t)$

or
$$x = \sin t\ f'(t) + \cos t\ f''(t). \quad ...(3)$$

Again, multiplying (1) by cos t and (2) by sin t and subtracting, we get

$$y = \cos t\ f'(t) - \sin t\ f''(t). \quad ...(4)$$

Now differentiating (3) and (4) w.r.t. t, we get

$dx/dt = \cos t\ f'(t) + \sin t\ f''(t) + \cos t\ f'''(t) - \sin t\ f''(t) = [f'(t) + f'''(t)]\cos t$

and $\qquad dy/dt = -[f'(t)\ t\ f'''(t)] \sin t.$

Now is s be the arc length in the direction of t increasing, then

$$\frac{ds}{dt} = \sqrt{\left\{\left(\frac{dx}{dt}\right)^2 + \left(\frac{dy}{dt}\right)^2\right\}}$$

$$= \sqrt{[\cos^2 t\,\{f'(t) + f'''(t)\}^2 + \sin^2 t\,\{f'(t) + f'''(t)\}^2]}$$

$$= [f'(t) + f'''(t)]\sqrt{(\cos^2 t + \sin^2 t)} = f'(t) + f'''(t).$$

Integrating both sides, we have $s = \int [f'(t) + f'''(t)]dt + c$

$= f(t) + f''(t) + c$, where c is the constant of integration.

Example 11:

Find the intrinsic equation of the cardioid $r = a(1 + \cos\theta)$, and hence, or otherwise, prove that $s^2 + 9\rho^2 = 16a^2$, where ρ is the radius of curvature at any point, and s is the length of the arc interceped between the vertex and the point.

Solution:

The given curve is $r = a(1 + \cos\theta)$. ...(1)

Differentiating (1) w.r.t. θ, we have $dr/d\theta = -a\sin\theta$.

$$\therefore \qquad \tan\phi = r\frac{d\theta}{dr} = \frac{r}{dr/d\theta}$$

$$= \frac{a(1+\cos\theta)}{-a\sin\theta}$$

$$= \frac{2\cos^2\frac{1}{2}\theta}{-2\sin\frac{1}{2}\theta\cos\frac{1}{2}\theta}$$

$$= -\cot\frac{1}{2}\theta = \tan\left(\frac{1}{2}\pi + \frac{1}{2}\theta\right).$$

Therefore $\qquad \phi = \frac{1}{2}\pi + \frac{1}{2}\theta$, so that

$$\psi = \theta + \phi = \theta + \frac{1}{2}\pi + \frac{1}{2}\theta = \frac{1}{2}\pi + \frac{3}{2}\theta$$

or $\qquad \frac{1}{2}\theta = \frac{1}{3}\left(\psi - \frac{1}{2}\pi\right).$ ...(2)

If s denotes the arc length of the cardioid measured from the vertex (*i.e.*, $\theta = 0$) to any point P (*i.e.*, $\theta = \theta$) in the direction of θ increasing, then

$$s = \int_0^\theta \sqrt{\left\{r^2 + \left(\frac{dr}{d\theta}\right)^2\right\}} d\theta = 2a\int_0^\theta \sqrt{\left\{(1+\cos\theta)^2 + \sin^2\theta\right\}} d\theta$$

$$= 2a\int_0^\theta \sqrt{\left\{1 + 2\cos\theta + \cos^2\theta + \sin^2\theta\right\}} d\theta$$

$$= 2a\int_0^\theta \sqrt{\left\{2(1+\cos\theta)\right\}} d\theta = 2a\int_0^\theta \cos\frac{1}{2}\theta d\theta$$

$$= 2a\left[2\sin\frac{1}{2}\theta\right]_0^\theta = 4a\sin\frac{1}{2}\theta. \qquad ...(3)$$

Eliminating θ between (2) and (3), we get

$$s = 4a\sin\left\{\frac{1}{3}\left(\psi - \frac{1}{2}\pi\right)\right\}, \qquad ...(4)$$

which is the required intrinsic equation.

Also $\rho = \dfrac{ds}{d\psi} = \dfrac{4a}{3}\cos\dfrac{1}{3}\left(\psi - \dfrac{1}{2}\pi\right)$, from (4)

or $$3\rho = 4a\cos\frac{1}{3}\left(\psi - \frac{1}{2}\pi\right). \qquad ...(5)$$

Squaring and adding (4) and (5), we get

$$s^2 + 9\rho^2 = (4a)^2\left\{\sin^2\frac{1}{3}\left(\psi - \frac{1}{2}\pi\right) + \cos^2\frac{1}{3}\left(\psi - \frac{1}{2}\pi\right)\right\}$$

$$= 16a^2 \times 1 = 16a^2.$$

Example 12:

Find the intrinsic equation of $r = a e^{\theta \cot\alpha}$, *where s is measured from the point (a, 0).*

Solution:

The given curve is $r = a e^{\theta \cot\alpha}$. ...(1)

Differentiating (1), w.r.t. θ, we have

$$(dr/d\theta) = a\cot\alpha . e^{\theta\cot\alpha} = r\cot\alpha.$$

We have $\tan\phi = r\dfrac{d\theta}{dr} = \dfrac{r}{dr/d\theta} = \dfrac{r}{r\cot\alpha} = \tan\alpha$

or $\phi = \alpha$ so that $\psi = \theta + \phi = \theta + \alpha$ or $\theta = \phi - \alpha$. ...(2)

If we measure the arc length s from the point $\theta = 0$ to any point P (r, θ) in the direction of increasing, we have

$$s = \int_0^\theta \sqrt{\left\{(dr / d\theta)^2 + r^2\right\}} d\theta$$

$$= \int_0^\theta \sqrt{\left\{r^2 \cot^2 \alpha + r^2\right\}} d\theta$$

$$= \int_0^\theta r \sqrt{\left(1 + \cot^2 \alpha\right)} d\theta = \operatorname{cosec} \alpha \int_0^\theta r d\theta$$

$$= \operatorname{cosec} \alpha \int_0^\theta a e^{\theta \cot \alpha} d\theta, \qquad [\because r = \alpha\, e^{\theta \cot \alpha}]$$

$$= a \operatorname{cosec} \alpha \left[\frac{e^{\theta \cot \alpha}}{\cot \alpha}\right]_0^\theta = a \sec \alpha\, [e^{\theta \cot \alpha} - 1]. \qquad ...(3)$$

Eliminating θ between (2) and (3), we get

$s = a \sec \alpha\, [e^{(\psi - a)\cot \alpha} - 1]$, which is the required intrinsic equating.

Example 13:

Find the arc length of the curve
$y = 1/2x^2 - 1/4\log x$ from $x = 1$ to $x = 2$.

Solution:

The given curve is $y = 1/2x^2 - 1/4\log x$. ...(1)

Differentiating (1) w.r.t. x, we get

$$\frac{dy}{dx} = x - \frac{1}{4x} = \frac{4x^2 - 1}{4x}.$$

$\therefore$ required length of the curve

$$= \int_1^2 \sqrt{\left\{1 + \left(\frac{dy}{dx}\right)^2\right\}} dx = \int_1^2 \sqrt{\left\{1 + \frac{(4x^2 - 1)^2}{16x^2}\right\}} dx$$

$$= \int_1^2 \frac{4x^2 + 1}{4x} dx = \int_1^2 \left(x + \frac{1}{4x}\right) dx$$

$$= \left[\frac{x^2}{2} + \frac{\log x}{4}\right]_1^2 = \frac{3}{2} + \frac{1}{4}\log 2.$$

Example 14:

Show that in the catenary $y = c \cosh (x/c)$, the length of arc from the vertex to any point is given by $s = c \sinh (x/c)$.

Solution:

The given catenary is $y = c \cosh (x/c)$.

The point A (0, c) is the vertex of the catenary and let P (x, y) be any point on it. We have to find the length or arc AP for which x varies from x = 0 to x = x.

Differentiating (1) w.r.t. x, we have

$$\frac{dy}{dx} = c \cdot \frac{1}{c} \sinh \frac{x}{c} = \sinh \frac{x}{c}.$$

If s denotes the arc length of the catenary measured in the direction of x increasing, then

$$\frac{ds}{dx} = \sqrt{\left\{1+\left(\frac{dy}{dx}\right)^2\right\}} = \sqrt{\left(1+\sinh^2 \frac{x}{c}\right)} = \cosh \frac{x}{c}$$

or $\qquad ds = \cosh (x/c)\, dx$. Integrating, we have

$$\int_0^s ds = \int_0^s \cosh \frac{x}{c} dx = \left[c \sinh \frac{x}{c}\right]_0^x = c \sinh \frac{x}{c}$$

or $\qquad s = c \sinh (x/c)$, which is the required arc length.

Example 15:

Show that in the parabola $\frac{2a}{r} = 1 + \cos\theta$,

$$\frac{ds}{d\psi} = \frac{2a}{\sin^3 \psi}.$$

Hence find the length of the arc intercepted between the vertex and an extremity of the latus rectum.

Solution:

The given equation of parabola is

$$\frac{2a}{r} = 1 + \cos\theta, \qquad ...(1)$$

in which the focus O is at pole.

From (1), we have

$$r = \frac{2a}{1+\cos\theta} = \frac{2a}{2\cos^2 \frac{1}{2}\theta} = a \sec^2 \frac{1}{2}\theta.$$

$$\therefore \quad \frac{dr}{d\theta} = 2a \sec \frac{1}{2}\theta \times \left(\sec \frac{1}{2}\theta \tan \frac{1}{2}\theta\right) \cdot \frac{1}{2}$$

$$= a \sec^2 \frac{1}{2}\theta \tan \frac{1}{2}\theta.$$

$$\therefore \qquad \left(\frac{ds}{d\theta}\right)^2 = r^2 + \left(\frac{dr}{d\theta}\right)^2$$

$$= a^2 \sec^4 \frac{1}{2}\theta + a^2 \sec^4 \frac{1}{2}\theta \tan^2 \frac{1}{2}\theta$$

$$= a^2 \sec^4 \frac{1}{2}\theta \left(1 + \tan^2 \frac{1}{2}\theta\right) = a^2 \sec^6 \frac{1}{2}\theta.$$

If s is the arc length of the parabola measured from the vertex A (*i.e.*, the point $\theta = 0$) to any point $P(r, \theta)$ in the direction of θ increasing, then

$$\frac{ds}{d\theta} = a \sec^3 \frac{1}{2}\theta.$$

We have $\qquad \cot\phi = \dfrac{1}{r}\dfrac{dr}{d\theta}$

$$= \frac{1}{a\sec^2 \frac{1}{2}\theta} a \sec^2 \frac{1}{2}\theta \tan \frac{1}{2}\theta$$

$$= \tan \frac{1}{2}\theta = \cot\left(\frac{1}{2}\pi - \frac{1}{2}\theta\right).$$

$$\therefore \qquad \phi = \frac{1}{2}\pi - \frac{1}{2}\theta.$$

$$\therefore \qquad \psi = \theta + \phi = \theta + \frac{1}{2}\pi - \frac{1}{2}\theta = \frac{1}{2}\pi + \frac{1}{2}\theta. \qquad \text{...(3)}$$

Now $\dfrac{ds}{d\psi} = \dfrac{ds}{d\theta}\cdot\dfrac{d\theta}{d\psi}$

$$= \left(a \sec^3 \frac{1}{2}\theta\right)\cdot 2, \qquad \left[\because \text{from(3)}, \frac{d\psi}{d\theta} = \frac{1}{2} \text{ so that } \frac{d\theta}{d\psi} = 2\right]$$

$$= 2a \sec^3\left(\psi - \frac{1}{2}\pi\right) = 2a \sec^3\left(\frac{1}{2}\pi - \psi\right)$$

$$= \frac{2a}{\cos^3\left(\frac{1}{2}\pi - \psi\right)} = \frac{2a}{\sin^3\psi}.$$

Now from $\dfrac{ds}{d\psi} = 2a \operatorname{cosec}^3 \psi$, we have

$$ds = 2a \operatorname{cosec}^3 \psi \, d\psi, \qquad \text{...(4)}$$

At the vertex A, $\theta = 0$ and at the extremity L of latus rectum LOL', $\theta = \pi/2$. So from (3), at A, $\psi = 1/2\pi$ and at L, $\psi = 3\pi/4$.

Integrating both sides of (4) from the point A to L, we have

$$\text{arc AL} = \int_{\pi/2}^{3\pi/4} 2a \operatorname{cosec}^3\psi \, d\psi$$

$$= 2a\int_{\pi/2}^{3\pi/4} \sqrt{(1+\cot^2\psi)}.\operatorname{cosec}^2\psi \, d\psi$$

$$= 2a\int_{0}^{-1} \sqrt{(1+t^2)}\cdot(-dt),$$

Putting $\cot\psi = t$, so that $-\operatorname{cosec}^2\psi \, d\psi = dt$

$$= 2a\int_{-1}^{0} \sqrt{(1+t^2)}dt$$

$$= 2a\left[\frac{t}{2}\sqrt{(+t^2)} + \frac{1}{2}\log\left\{t+\sqrt{(1+t^2)}\right\}\right]_{-1}^{0}$$

$$= 2a\left[0+\frac{1}{2}\log 1 - \left\{-\frac{1}{2}\sqrt{2}+\frac{1}{2}\log\left(-1+\sqrt{2}\right)\right\}\right]$$

$$= 2a\left[\frac{1}{2}\sqrt{2}-\frac{1}{2}\log\left(\sqrt{2}-1\right)\right]$$

$$= 2a\left[\frac{1}{2}\sqrt{2}+\frac{1}{2}\log\left(\frac{1}{\sqrt{2}-1}\right)\right]$$

$$= 2a\left[\frac{1}{2}\sqrt{2}+\frac{1}{2}\log\left(\sqrt{2}+1\right)\right]$$

$$= a\left[\sqrt{2} + \log(\sqrt{2}+1)\right].$$

Example 16:

Find the intrinsic equation of the spirals $r = a\theta$, the arc being measured from the pole.

Solution:

The given curve is $r = a\theta$. ...(1)

Differentiating (1) w.r.t. θ,

we have $dr/d\theta = a$.

$$\text{Therefore } \tan\phi = r\frac{d\theta}{dr} = \frac{r}{dr/d\theta}$$

$$= \frac{a\theta}{a} = \theta.$$

$\therefore \quad \phi = \tan^{-1}\theta$

so that $\psi = \theta + \phi = \theta + \tan^{-1}\theta$. ...(2)

If s denotes the arc length of the spiral measured from the pole (0, 0) to any point P (r, θ), then

$$s = \int_0^{\theta} \sqrt{\left\{r^2 + \left(\frac{dr}{d\theta}\right)^2\right\}} d\theta$$

$$= \int_0^{\theta} \sqrt{\left(a^2\theta^2 + a^2\right)} d\theta$$

$$= a\int_0^{\theta} \sqrt{\left(\theta^2 + 1\right)} d\theta$$

$$= a\left[\frac{\theta}{2}\sqrt{\left(\theta^2 + 1\right)} + \frac{1}{2}\log\left\{\theta + \sqrt{\left(\theta^2 + 1\right)}\right\}\right]_0^{\theta}$$

$$= \frac{1}{2}a\ [\theta\ \sqrt{(1 + \theta^2)} + \log\ \{\theta + \sqrt{(1 + \theta^2)}. \quad ...(3)$$

The required intrinsic equation is obtained by eliminating θ between (2) nad (3).

Example 17:

Find the intrinsic equation of the equiangular spiral $p = r \sin \alpha$.

Solution:

The given pedal equation of the curve is $p = r \sin \alpha$. ...(1)

Differentiating (1) w.r.t. r, we have

$$d\pi/dr = \sin \alpha.$$

$$\therefore \qquad \rho = \frac{ds}{d\psi} = r\frac{dr}{dp} = \frac{r}{dp/dr} = \frac{r}{\sin\alpha} = r \operatorname{cosec} \alpha. \quad ...(2)$$

If we measure the arc length s from the point r = 0 in the direction of r increasing, we have

$$s = \int_0^r \frac{r\,dr}{\sqrt{(r^2 - p^2)}} = \int_0^r \frac{r\,dr}{\sqrt{(r^2 - r^2\sin^2\alpha)}} = \int_0^r \sec\alpha\, dr$$

$$= \sec\alpha \int_0^r dr = \sec\alpha\ [r]_0^r = r\sec\alpha. \quad ...(3)$$

Eliminating r between (2) and (3), we have

$$\frac{(ds/d\psi)}{s} = \frac{\operatorname{cosec}\alpha}{\sec\alpha} = \cot\alpha. \qquad \text{[dividing (2) by (3)]}$$

or $$ds/s = \cot\alpha\, d\psi.$$

Integrating, $\log s = \psi \cot \alpha + \log \alpha$, where a is constant of integration

or $\log (s/a) = \psi \cot \alpha$ or $s = a\,e^{\psi \cot \alpha}$, which is the required intrinsic equation of the curve.

Example 18:

Find the intrinsic equation of the curve $p^2 = r^2 - a^2$.

Solution:

The given curve is $p^2 = r^2 - a^2$ ...(1)

Differentiating (1) w.r.t. r, we have

$$2p\,(dp/dr) = 2r \text{ or } r\,(dr/dp) = p.$$

$$\therefore \qquad \rho = \frac{ds}{d\psi} = r\frac{dr}{dp} = \pi = \sqrt{(r^2 - a^2)}, \text{ [from (1)].} \qquad ...(2)$$

Also from the equation of the curve we have p = 0 for r = a.

If we measure the arc length s (from r = a) in the direction of r incursion, we have

$$s = \int_a^r \frac{r\,dr}{\sqrt{(r^2 - p^2)}} = \int_a^r \frac{r\,dr}{a}, \qquad [\because r^2 - p^2 = a^2]$$

$$= \frac{1}{a}\left[\frac{r^2}{2}\right]_a^r = \frac{1}{2a}\left[r^2 - a^2\right]$$

or $\qquad 2as = r^2 - a^2$ or $\sqrt{(2as)} = \sqrt{(r^2 - a^2)}$ ...(3)

Eliminating r between (2) and (3), we have

$$\frac{ds}{d\psi} = \sqrt{(2as)} \text{ or } \frac{ds}{\sqrt{s}} = \sqrt{(2a)}\, dy.$$

If s = 0 when $\psi = 0$, then integrating, we have

$$\int_0^s \frac{ds}{\sqrt{s}} = \sqrt{(2a)}\int_0^\psi d\psi.$$

$\therefore\ 2\sqrt{s} = \sqrt{(2a)}\psi$ or $s = 1/2a\psi^2$, which is the required intrinsic equation.

Example 19:

Find the intrinsic equation of the curve for which the length of the arc measured from the origin varies as the square root of the ordinate. Find also parametric equation of the curve in terms of any parameter.

Solution:

Let s denote the arc length of the curve measured from the origin to any point P (x, y) such that s increases as y increases. As given $s \propto \sqrt{y}$ so that $s = \lambda\sqrt{y}$, where λ is some constant.

Choosing this constant $\lambda = \sqrt{(8a)}$ (Note), we have

$$s = \sqrt{(8ay)} \text{ or } s^2 = 8ay. \qquad ...(1)$$

Now differentiating (1) w.r.t. y, we have

$$2s\,(ds/dy) = 8a \text{ or } ds/dy = 4a/s. \qquad ...(2)$$

Now we know that $dy/ds = \sin\psi$.

$\therefore \qquad \sin\psi = dy/ds = s/4a,$ [from (2)]

or $\qquad s = 4a \sin\psi$, which is the required intrinsic equation.

Again from (1), we have

$$y = \frac{s^2}{8a} = \frac{16a^2 \sin^2\psi}{8a}$$
$$= a\,(1 - \cos^2\psi),\ [\because s = 4a\sin\psi]. \qquad ...(3)$$

Also $\qquad \dfrac{ds}{dx} = \dfrac{ds}{d\psi}\cdot\dfrac{d\psi}{dx} = 4a\cos\psi\,\dfrac{d\psi}{dx}, \qquad \left[\because \dfrac{ds}{d\psi} = 4a\cos\psi\right]$

or $\qquad \dfrac{1}{\cos\psi} = 4a\cos\psi\,\dfrac{d\psi}{ds}, \qquad \left[\because \dfrac{dx}{ds} = \cos\psi\right]$

or $\qquad dx = 4a\cos^2\psi\,d\psi = 2a\,(1 + \cos 2\psi)\,d\psi. \qquad ...(4)$

If $x = 0$ when $\psi = 0$, then integrating (4), we get

$$\int_0^x dx = 2a\int_0^{\psi} (1 + \cos 2\psi)\,d\psi$$

or $\qquad x = 2a\left[\psi + \dfrac{1}{2}\sin 2\psi\right]_0^{\psi}$

or $\qquad x = a\,[2\psi + \sin^2\psi]. \quad ...(5)$

So from (3) and (5), the required parametric equation of the curve are $x = a\,(2\psi + \sin^2\psi)$ and $\psi = a\,(1 - \cos 2\psi)$, which are the parametric equation of a cycloid.

Example 20:

Find the cartesian equation of the curve whose intricsic equation is $s = c\tan\psi$ when it is given that at $\psi = 0$, $x = 0$ and $y = c$.

Solution:

The given intrinsic equation of the curve is $s = c\tan\psi$. ...(1)

Differentiating (1) w.r.t. 'x', we have

$$ds/dx = c\sec^2\psi.\,(dy/dx) \qquad ...(2)$$

Also $\qquad \dfrac{ds}{dx} = \sqrt{\left\{1 + \left(\dfrac{dy}{dx}\right)^2\right\}} = \sqrt{(1 + \tan^2\psi)} = \sec\psi. \qquad ...(3)$

Equating the values of ds/dx from (2) and (3), we get

$$c\sec^2\psi.(d\psi/dx) = \sec\psi \text{ or } dx = c\sec\psi\,d\psi. \qquad ...(4)$$

Integrating both sides of (1), we get

$x + A = c \log (\sec \psi + \tan \psi)$, where A is constant of integration.

But as given $\psi = 0$ when $x = 0$ so that $A = 0$.

Therefore $x = c \log (\sec \psi + \tan \psi)$

or $$e^{x/c} = \sec \psi + \tan \psi. \quad ...(5)$$

Now $$e^{-x/c} = \frac{1}{e^{x/c}} = \frac{1}{\sec \psi + \tan \psi}$$

$$= \frac{\sec \psi - \tan \psi}{\sec^2 \psi - \tan^2 \psi}$$

$$= \sec \psi - \tan \psi. \quad ...(6)$$

Subtracting (6) from (5), we get

$$e^{x/c} - e^{-x/c} = 2 \tan \psi$$

or $$\tan \psi = \frac{e^{x/c} - e^{-x/c}}{2}$$

or $$\frac{dy}{dx} = \sinh \frac{x}{c}$$

or $$dy = \sin h\ (x/c)\ dx.$$

Integrating both sides, we get

$$y + B = c \cosh (x/c). \quad ...(7)$$

But (as given) when $x = 0$,

$y = c$ so that $B = 0$.

Therefore putting $B = 0$ in (7), we get

$y = c \cosh (x/c)$, which is the required cartesian equation of the given curve.

Example 21:

Show that the length of the curve $y = \log \sec x$ between the points where $x = 0$ and $x = 1/3\pi$ is $\log (2 + \sqrt{3})$.

Solution:

The given curve is $y = \log \sec x$. ...(1)

Differentiating (1) w.r.t. x, we get

$$\frac{dy}{dx} = \frac{1}{\sec x} \sec x \tan x = \tan x.$$

Now $$\left(\frac{ds}{dx}\right)^2 = 1 + \left(\frac{dy}{dx}\right)^2 = 1 + \tan^2 x = \sec^2 x. \quad ...(2)$$

If the arc length s of the given curve is measured from x = 0 in the direction of x increasing, we have

$$\frac{ds}{dx} = \sec x \text{ or } ds = \sec x\, dx.$$

Therefore if s_1 denotes the arc length from x = 0 to $x = \frac{1}{3}\pi$, then

$$\int_0^{s_1} ds = \int_0^{\pi/3} \sec x\, dx = [\log(\sec x + \tan x)]_0^{\pi/3}$$

or

$$s_1 = \left[\log\left(\sec\frac{1}{3}\pi + \tan\frac{1}{3}\pi\right) - \log 1\right] = \log(2 + \sqrt{3}).$$

Example 22:

Find the length of the curve $y = \log[(e^x - 1)/(e^x + 1)]$ from x = 1 to x = 2.

Solution:

The given curve is $y = \log[(e^x - 1)/(e^x + 1)]$

or

$$y = \log(e^x - 1) - \log(e^x + 1). \qquad ...(1)$$

Differentiating (1) w.r.t. x, we get

$$\frac{dy}{dx} = \frac{e^x}{e^x - 1} - \frac{e^x}{e^x + 1} = \frac{2e^x}{(e^{2x} - 1)}.$$

Now

$$\left(\frac{ds}{dx}\right)^2 = 1 + \left(\frac{dy}{dx}\right)^2 = 1 + \left\{\frac{2e^x}{e^{2x} - 1}\right\}^2 = 1 + \frac{4e^{2x}}{(e^{2x} - 1)^2}$$

$$= \frac{(e^{2x} - 1)^2 + 4e^{2x}}{(e^{2x} - 1)^2} = \frac{(e^{2x} + 1)^2}{(e^{2x} - 1)^2}.$$

Measuring the arc length s in the direction of x increasing, we have

$$\frac{ds}{dx} = \frac{e^{2x} + 1}{e^{2x} - 1}$$

$$= \frac{e^x + e^{-x}}{e^x - e^{-x}}, \text{ dividing the Nr. and the Dr. by } e^x$$

or

$$ds = \frac{e^x + e^{-x}}{e^x - e^{-x}} dx.$$

$\therefore$ the required length s_1 is given by

$$s_1 = \int_1^2 \frac{e^x + e^{-x}}{e^x - e^{-x}} dx$$

$$= \left[\log(e^x - e^{-x}\right]_1^2,$$

numerator being the diff. coeff. of denominator

$= \log (e^2 - e^{-2}) - \log (e - e^{-1})$

$= \log [\{e^2 - (1/e^2)\}/\{e - (1/e)\}]$

$= \log (e + 1/e).$

Example 23:

If s be the length of the arc of the catenary. $y = c \cosh (x/c)$ from the vertex (0, c) to the point (x, y), show that $s^2 = y^2 - c^2$.

Solution:

The given curve is $y = c \cosh (x/c)$. ...(1)

The length s of arc extending from the vertex (0, c) to any point (x, y) is given by

$$s = c \sinh (x/c) \qquad ...(2)$$

Squaring and subtracting (2) from (1),

we get $y^2 - s^2 = c^2 \cosh^2(x/c) - c^2 \sinh^2 (x/c) = c^2$

or $\qquad y^2 - c^2 = s^2.$

This was to be proved.

Example 24:

If A denotes the area between the curve $y = c \cosh (x/c)$ and the two ordinates $x = x_1$ and $x = x_2$ and the x-axis and 's' stands for the length of the intervening arc, then prove that $A = cs$.

Or

In the catenary $y = c \cosh (x/c)$, prove that the area between the curve, the axis of x and the ordinates of two points on the curve, varies as the length of the intervening curve.

Solution:

The required area

$$A = \int_{x_1}^{x_2} y\,dx = \int_{x_1}^{x_2} c \cosh\left(\frac{x}{c}\right) dx$$

$$= c \int_{x_1}^{x_2} \cosh \frac{x}{c}\,dx. \qquad ...(1)$$

The required arc length

$$s = \int_{x_1}^{x_2} \sqrt{\left\{1 + \left(\frac{dy}{dx}\right)^2\right\}}\,dx$$

$$= \int_{x_1}^{x_2} \sqrt{\left(1 + \sinh^2 \frac{x}{c}\right)} dx$$

$$= \int_{x_1}^{x_2} \cosh \frac{x}{c} dx. \qquad \ldots(2)$$

From (1) and (2), we observe that A = cs.

Example 25:

Find the whole length of the curve (astroid) $x = a \cos^3 t$, $y = a \sin^3 t$.

Solution:

The given parametric equations of the astroid are

$$x = a \cos^3 t,\ y = a \sin^3 t. \qquad \ldots(1)$$

We have $dx/dt = -3a \cos^2 t \sin t$,

$$dy\ dt = 3a \sin^2 t \cos t.$$

The astroid is symmetrical about both the axes. For the arc of the astroid lying in the first quadrant, we have t = 0 at the point (a, 0) and t = π/2 at the point (0, a).

Now
$$\left(\frac{ds}{dt}\right)^2 = \left(\frac{dx}{dt}\right)^2 + \left(\frac{dy}{dt}\right)^2$$

$$= 9a^2 \cos^4 t \sin^2 t + 9a^2 \sin^4 t \cos^2 t$$

$$= 9a^2 \sin^2 t \cos^2 t (\sin^2 t + \cos^2 t)$$

$$= (3a \sin t \cos t)^2. \qquad \ldots(2)$$

If s denotes the arc length of the astroid measured from the point t = 0 to any point P towards the point t = π/2, then s increases as t the increases. Therefore ds/dt will be taken with positive sign. So taking square root of both sides of (2), we have $ds/dt = 3a \sin t \cos t$ or $ds = 3a \sin t \cos t\ dt$.

Let s_1 denote the arc length of the astroid lying in the first quadrant. Then

$$\int_0^{s_1} ds = \int_0^{\pi/2} 3a \sin t \cos t\ dt \text{ or } s_1 = 3a \left[\frac{\sin^2 t}{2}\right]_0^{\pi/2} = \frac{3a}{2}.$$

Whole length of the curve

= 4 × length of the curve lying in the Ist quadrant = 4×(3a/2) = 6a.

Example 26:

Find the length of the curve

$x = \dfrac{c^2}{a}\cos^3 t,\ y = \dfrac{c^2}{a}\sin^3 t$ *which is the evolute of an ellipse.*

Solution:

Do your self.

Replace a by c^2/a.

The required length = 6 (c^2/a) = $6c^2/a$.

Example 27:

Find the whole length of the curve (Hypocycloid) $x = a\cos^3 t,\ y = b\sin^3 t$.

Solution:

The curve is symmetrical about both the axes and in the first quadrant t varies from 0 to $1/2\pi$.

Here $dx/dt = -\,3a\cos^2 t\sin t,\ dy/dt = 3b\sin^2 t\cos t$.

$\therefore$ the required whole length of the curve

$$= 4 \times \text{length of the curve in the first quadrant}$$

$$= 4\int_0^{\pi/2}\sqrt{\left\{\left(\frac{dx}{dt}\right)^2 + \left(\frac{dy}{dt}\right)^2\right\}}dt$$

$$= 4\int_0^{\pi/2}\sqrt{\{9a^2\cos^4 t\sin^2 t + 9b^2\sin^4 t\cos^2 t\}}\ dt$$

$$= 4\int_0^{\pi/2} 3\sin t\cos t\ \sqrt{(a^2\cos^2 t + b^2\sin^2 t)}\ dt.$$

Now put $a^2\cos^2 t + b^2\sin^2 t = z^2$,

so that $(-2a^2\sin t\cos t + 2b^2\sin t\cos t)\ dt = 2z\,dz$.

$\therefore$ $\sin t\cos t\,dt = (z/(b^2 - a^2)\}\ dz$.

Also $z = a$ when $t = 0$ and $z = b$ when $t = \pi/2$.

Hence the required length

$$= 12\int_a^b z\cdot\frac{zdz}{b^2 - a^2} = \frac{12}{b^2 - a^2}\int_a^b z^2 dz$$

$$= \frac{12}{b^2 - a^2}\left[\frac{z^3}{3}\right]_a^b = 4\ \frac{b^3 - a^3}{b^2 - a^2}$$

$$= 4 \times \frac{(b^2 + ab + a^2)}{b + a}.$$

Example 28:

Find the length of the arc of the parabola $y^2 = 4ax$ extending from the vertex to an extremity of the latus rectum.

Solution:

The given equation of parabola is

$$y^2 = 4ax. \qquad ...(1)$$

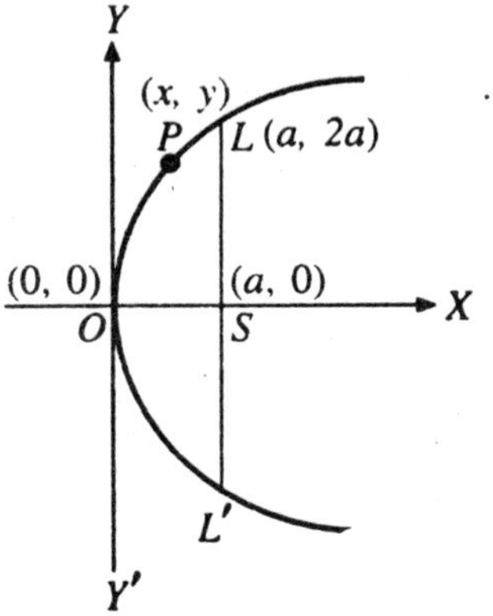

The point O (0, 0) is the vertex of the parabola and the point L (a, 2a) is an extremity of the latus rectum LSL'. We have to find the length of arc OL. Defferentiating (1) w.r.t. x, we get 2y (dy/dx) = 4a.

$$\therefore \qquad dy/dx = 2a/y$$

or

$$dx/dy = y/2a.$$

Fig. 2.3

Now
$$\left(\frac{ds}{dy}\right)^2 = 1 + \left(\frac{dx}{dy}\right)^2$$

$$= 1 + \frac{y^2}{4a^2} = \frac{1}{4a^2}(4a^2 + y^2). \qquad ...(2)$$

If 's' denotes the length of the parabola measured from the vertex O to any point P (x, y) towards the point L, then s increases as y increases. Therefore, ds/dy will be positive So extracting the square root of (2) and keeping the positive sign, we have

$$\frac{ds}{dy} = \frac{1}{2a}\sqrt{(4a^2 + y^2)},$$

or
$$ds = \frac{1}{2a}\sqrt{(4a^2 + y^2)}dy.$$

Let s_1 denote the arc length OL. Then

$$\int_0^{s_1} ds = \int_0^{2a} \frac{1}{2a}\sqrt{(4a^2 + y^2)}dy$$

or
$$s_1 = \frac{1}{2a}\left[\frac{y}{2}\sqrt{(4a^2 + y^2)} + \frac{4a^2}{2}\log\{y + \sqrt{(4a^2 + y^2)}\}\right]_0^{2a}$$

$$= \frac{1}{2a}[a\sqrt{(4a^2 + y^2)} + 2a^2 \log\left\{2a + \sqrt{(8a^2)}\right\}$$

$$= \frac{1}{2a}[2\sqrt{2a^2} + 2a^2 \log\{(2a + 2\sqrt{(8a^2)}\} - 0 - 2a^2 \log(2a)]$$

$$= \frac{1}{2a}[2\sqrt{2a^2} + 2a^2 \log\{(2a + 2\sqrt{2}a)/2a\}]$$

$$= \frac{2a^2}{2a}[\sqrt{2} + \log(1 + \sqrt{2})] = a[\sqrt{2} + \log(1 + \sqrt{2})].$$

Example 29:

Find the length of the arc of the parabola $y^2 = 4ax$ cut off by its latus rectum.

Solution:

Do your self.

Here we have to find the arc length L'OL.

Example 30:

Find the length of an arc of the parabola measured from the vertex.

Solution:

Let the given parabola by $y^2 = 4ax$. ..(1)

Differentiating (1) w.r.t. x, we get $2y\,(dy/dx) = 4a$.

$\therefore$ $dy/dx = 2a/y$ or $dx/dy = y/2a$.

If s denotes the arc length of the parabola $y^2 = 4ax$ measured from the vertex O to any point P (x, y) lying on the upper half of the parabola, then s increases as y increases.

$\therefore$ the required length of are

$$s = \int_0^y \sqrt{\left\{1 + \left(\frac{dx}{dy}\right)^2\right\}}dy$$

$$= \int_0^y \sqrt{\left(1 + \frac{y^2}{4a^2}\right)}dy = \frac{1}{2a}\int_0^y \sqrt{(y^2 + 4a^2)}dy$$

$$= \frac{1}{2a}\left[\frac{y}{2}\sqrt{(y^2 + 4a^2)} + \frac{4a^2}{2}\log\{y + \sqrt{(y^2 + 4a^2)}\}\right]_0^y$$

$$= \frac{1}{4a}[y\sqrt{(y^2 + 4a^2)} + 4a^2 \log\{y + \sqrt{(y^2 + 4a^2)}\} - 4a^2 \log 2a]$$

$$= \frac{1}{4a}\left[y\sqrt{(y^2+4a^2)}+4a^2\log\left\{\frac{y+\sqrt{(y^2+4a^2)}}{2a}\right\}\right].$$

Example 31:

Find the length of the are of the parabola $y^2 = 4ax$ cut off by the line $y = 3x$.

Solution:

We have $$y^2 = 4ax \quad ...(1)$$

and $$y = 3x. \quad ...(2)$$

Substituting for x in (1) from (2), we get

$y^2 = 4a.\ (y/3)$ or $3y^2 - 4ay = 0$ giving $y = 0,\ 4a/3$.

$\therefore$ from (2) the corresponding values of x are 0 and 4a/9. Hence, the points of intersection of the parabola and the given line are (0, 0) and (4a/9, 4a/3).

Also differentiating (1) w.r.t. x, we get 2y (dy/dx) = 4a.

$\therefore$ $$dy/dx = 2a/y \text{ or } dx/dy = y/2a.$$

If s denotes the arc length of the parabola measured from the point (0, 0) to the point (4a/9, 4a/3), then

$$s = \int_0^{4a/3}\sqrt{\left[1+\left(\frac{dx}{dy}\right)^2\right]}dy = \int_0^{4a/3}\sqrt{\left[1+\frac{y^2}{4a^2}\right]}dy$$

$$= \frac{1}{2a}\int_0^{4a/3}\sqrt{(y^2+4a^2)}dy$$

$$= \frac{1}{2a}\left[\frac{y}{2}\sqrt{(y^2+4a^2)}+\frac{4a^2}{2}\log\{y+\sqrt{(y^2+4a^2)}\}\right]_0^{4a/3}$$

$$= (1/2a)\left[\frac{1}{2}\cdot\frac{4}{3}a\sqrt{\left(\frac{16}{9}a^2+4a^2\right)}\right.$$

$$\left.+2a^2\log\left\{\frac{4}{3}a+\sqrt{\left(\frac{16}{9}a^2+4a^2\right)}\right\}-2a^2\log(2a)\right]$$

$$= (1/2a)\left[\frac{2}{3}a\cdot\frac{2}{3}a\sqrt{(13)}+2a^2\log\left\{\frac{4}{3}a+\frac{2}{3}a\sqrt{(13)}\right\}-2a^2\log(2a)\right]$$

$$= a\left[\frac{2}{9}\sqrt{(13)}+\log\left\{\frac{2}{3}+\frac{1}{3}\sqrt{(13)}\right\}\right]$$

$$= a\left[\frac{2}{9}\sqrt{(13)} + \log\left\{\frac{2+\sqrt{13})}{3}\right\}\right]$$

Example 32:

Show that the length of the arc of the parabola $y^2 = 4ax$ which is intercepted between the point of intersection of the parabola and the straight line $3y = 8x$ is $a\left(\log 2 + \frac{15}{16}\right)$.

Solution:

Solving $\quad y^2 = 4ax,$

and $\quad 3y = 8x$ for y,

we get $\quad y^2 = 4a.\ (3y/8),$

or $\quad 2y^2 - 3ay = 0,$

or $\quad y = 0,\ 3a/2.$

Thus, the parabola and the straight line intersect at the points where $y = 0$ and $3a/2$. We need not find the x-coordinates of these points.

Also differentiating $y^2 = 4ax$, we get

$$2y\ (dy/dx) = 4a \text{ or } (dy/dx) = (2a/y) \text{ or } (dx/dy) = y/2a.$$

∴ If s denotes the arc length of the parabola measured in the direction of y increasing, then

$$\frac{ds}{dy} = \sqrt{\left\{1+\left(\frac{dx}{dy}\right)^2\right\}} = \frac{1}{2a}\sqrt{(4a^2+y^2)},$$

or
$$ds = \frac{1}{2a}\sqrt{(4a^2+y^2)}dy.$$

Let s_1 denote the required arc length from $y = 0$ to $y = 3a/2$. Then

$$\int_0^{s_1} ds = \int_0^{3a/2} \frac{1}{2a}\sqrt{4a^2+y^2)}dy$$

or
$$s_1 = \frac{1}{2a}\left[\frac{y}{2}\sqrt{(4a^2+y^2)} + \frac{4a^2}{2}\log\{y+\sqrt{(4a^2+y^2)}\}\right]_0^{3a/2}$$

$$= \frac{1}{4a}\left[\frac{3a}{2}\sqrt{\left(\frac{25a^2}{4}\right)} + 4a^2\log\left\{\frac{3a}{2}+\sqrt{\left(\frac{25a^2}{4}\right)}\right\} - 4a^2\log 2a\right]$$

$$= a\left[\frac{15}{16} + \log 2\right].$$

Example 33:

Find the length of the arc of the parabola $x^2 = 4ay$ from the vertex to an extremity of the latus rectum.

Solution:

The given parabola is $x^2 = 4ay$, ...(1)

whose vertex is the point (0, 0) and whose axis is along the y-axis.

Let s_1 denote the arc length of the parabola measured from the vertex O (0, 0) to an extremity of the latus rectum (2a, a).

Differentiating (1) w.r.t. x, we get $2x = 4a\,(dy/dx)$ or $(dy/dx) = x/2a$.

$$\therefore \quad \text{the required arc length } s_1 = \int_0^{2a} \sqrt{\left\{1+\left(\frac{dy}{dx}\right)^2\right\}}dx$$

$$= \int_0^{2a} \sqrt{\left\{1+\frac{x^2}{4a^2}\right\}}dx = \frac{1}{2a}\int_0^{2a} \sqrt{(x^2+4a^2)}\,dx$$

$$= \frac{1}{2a}\left[\frac{x}{2}\sqrt{(x^2+4a^2)}+\frac{4a^2}{2}\log\{x+\sqrt{(x^2+4a^2)}\}\right]_0^{2a}$$

$$= a\,[\sqrt{2} + \log(1+\sqrt{2})].$$

Example 34:

Prove that the lengths of the loop of the curve $3ay^2 = x(x-a)^2$ is $4a/\sqrt{3}$.

Solution:

The given curve is symmetric about the x-axis.

At $y = 0$, we have $x = 0$ and $x = a$. So for the loop, x varies from 0 to a. The equation of the given curve is $3ay^2 = x(x-a)^2$.

Taking logarithm, we gave $\log 3a + 2\log y = \log x + 2\log(x-a)$.

Now differentiating w.r.t. x, we get

$$\frac{2}{y}\frac{dy}{dx} = \frac{1}{x}+\frac{2}{x-a} = \frac{3x-a}{x(x-a)}.$$

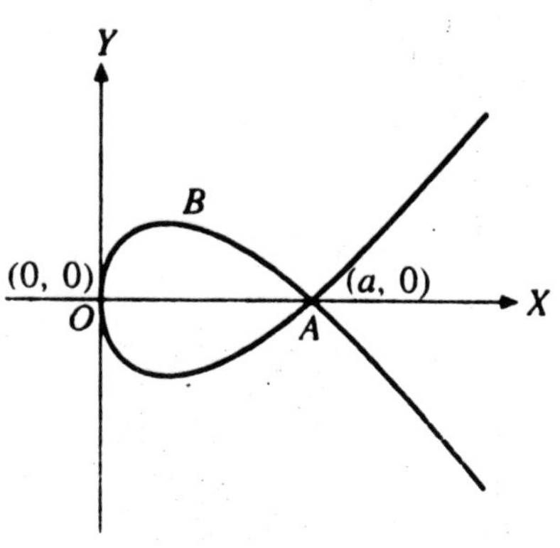

Fig. 2.4

Thus $\left(\frac{dy}{dx}\right)^2 = \frac{(3x-a)^2}{x^2(x-a)^2} \cdot \frac{y^2}{4}$

$$= \frac{(3x-a)^2 x(x-a)^2}{x^2(x-a)^2 \cdot 12a} = \frac{(3x-a)^2}{12ax}.$$

$$\therefore \quad \frac{ds}{dx} = \sqrt{\left\{1+\left(\frac{dy}{dx}\right)^2\right\}}$$

$$= \sqrt{\left\{1+\frac{(3x-a^2)}{12ax}\right\}} = \frac{3x+a}{\sqrt{(12ax)}}.$$

$\therefore$ the required length of the whole loop

$$= 2\int_0^a \sqrt{\left\{1+\left(\frac{dy}{dx}\right)^2\right\}}dx$$

$$= \int_0^a \frac{3x+a}{\sqrt{(12ax)}}dx$$

$$= \frac{1}{\sqrt{(3a)}}\int_0^a (3x^{1/2}+ax^{-1/2})\,dx$$

$$= \frac{1}{\sqrt{(3a)}}\left[3 \cdot \frac{2}{3}x^{3/2}+2ax^{1/2}\right]_0^a$$

$$= \frac{1}{\sqrt{(3a)}}[4a^{3/2}] = \frac{4a}{\sqrt{3}}.$$

Example 35:

Find the perimeter of the loop of the curve $9ay^2 = (x-2a)(x-5a)^2$.

Solution:

The given equation of the curve is $9ay^2 = (x-2a)(x-5a)^2$. ...(1)

Shifting the origin to the point (2a, 0), the equation (1) becomes $9ay^2 = x(x-3a)^2$. (Here we have 3a in place of a). The required length is

$$= 2\int_0^{3a} \frac{x+a}{\sqrt{(4ax)}}dx = 4a\sqrt{3}.$$

Example 15:

Find the perimeter of the loop of the curve $3ay^2 = x^2(a-x)$.

Solution:

The given curve is $3ay^2 = x^2 (a - x)$ ...(1)

Her., ıne curve is symmetrical about the x-axis. Putting $y = 0$, we get $x = 0$, $x = a$. So the loop lies between $x = 0$ and $x = a$. Differentiating (1) w.r.t. x, we get

$$6ay \frac{dy}{dx} = 2ax - 3x^2$$

or $$\frac{dy}{dx} = \frac{x(2a - 3x)}{6ay}.$$

$$\therefore \quad 1 + \left(\frac{dy}{dx}\right)^2 = 1 + \frac{x^2 (2a - 3x)^2}{36a^2 y^2} = 1 + \frac{x^2 (2a - 3x)^2}{12a\, x^2 (a - x)},$$

substituting for $3ay^2$ from (1)

$$= 1 + \frac{(2a - 3x)^2}{12a\,(a - x)} = \frac{12a^2 - 12ax + (2a - 3x)^2}{12a(a - x)} = \frac{(4a - 3x)^2}{12a(a - x)}$$

$\therefore$ the required length of the loop

= twice the length of the half loop of lying above the x-axis, (by symmetry)

$$= 2\int_0^a \sqrt{\left\{1 + \left(\frac{dy}{dx}\right)^2\right\}}dx = 2\int_0^a \sqrt{\left\{\frac{(4a - 3x)^2}{12a(a - x)}\right\}}dx$$

$$= \frac{1}{\sqrt{(3a)}}\int_0^a \frac{(4a - 3x)}{\sqrt{(a - x)}}dx$$

$$= \frac{1}{\sqrt{(3a)}}\int_0^a \frac{3(a - x) + a}{\sqrt{(a - x)}}dx$$

$$= \frac{1}{\sqrt{(3a)}}\int_0^a \left[\frac{3(a - x)}{\sqrt{(a - x)}} + \frac{a}{\sqrt{(a - x)}}\right]dx$$

$$= \frac{1}{\sqrt{(3a)}}\int_0^a [3\sqrt{(a - x)} + a(a - x)^{-1/2}]dx$$

$$= \frac{1}{\sqrt{(3a)}}\left[-3 \cdot \frac{2}{3}(a - x)^{3/2} - a \cdot 2(a - x)^{1/2}\right]_0^a$$

$$= \frac{1}{\sqrt{(3a)}}\left[2a^{3/2} + 2a^{3/2}\right] = \frac{4a}{\sqrt{3}}.$$

Example 36:

Show that the whole length of the curve $x^2 (a^2 - x^2) = 8a^2y^2$ is $\pi a \sqrt{2}$.

Solution:

The given curve is $x^2 (a^2 - x^2) = 8a^2y^2$. ...(1)

The curve (1) is symmetrical about both axes and it passes through the origin. Putting $y = 0$ in the given equation of the curve, we get

$$x^2 (a^2 - x^2) = 0$$

or
$$x = 0,$$
$$x = \pm a.$$

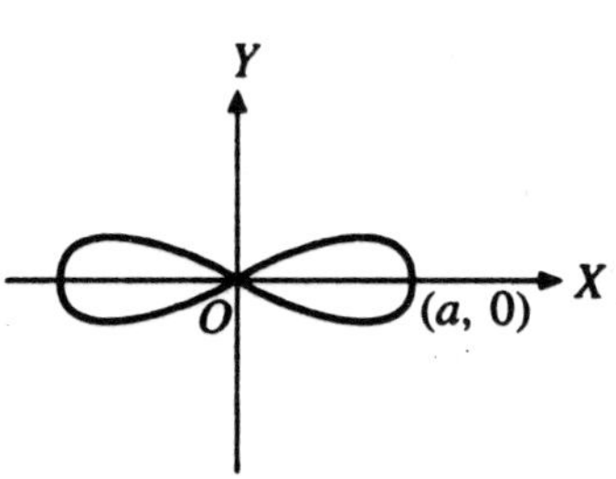

Fig. 2.5

So the curve passes through the points (0, 0), (a, 0) and (–a, 0). Therefore for one loop x varies form 0 to a.

$\therefore$ the required whole length of the curve

= 4 × length of the half loop (lying above x-axis)

$$= 4\int_0^a \sqrt{\left\{1 + \left(\frac{dy}{dx}\right)^2\right\}}dx.$$

Nwo differentiating (1) w.r.t. x, we get

$$16a^2 y \frac{dy}{dx} = 2a^2 x - 4x^3 \text{ or } \frac{dy}{dx} = \frac{(a^2 - 2x^2)}{8a^2 y} x.$$

$$\therefore \quad 1 + \left(\frac{dy}{dx}\right)^2 = 1 + \frac{x^2 (a^2 - 2x^2)^2}{64a^4 y^2}$$

$$= 1 + \frac{x^2 (a^2 - 2x^2)^2}{8a^2 x^2 (a^2 - x^2)},$$ substituting for $8a^2y^2$ form (1)

$$= 1 + \frac{(a^2 - 2x^2)^2}{8a^2 (a^2 - x^2)} = \frac{(3a^2 - 2x^2)^2}{8a^2 (a^2 - x^2)}.$$

$\therefore$ From (2), the required whole length of the curve

$$= 4\int_0^a \sqrt{\left\{\frac{3a^2 - 2x^2)^2}{8a^2(a^2 - x^2)}\right\}}dx$$

$$= \frac{2}{a\sqrt{2}}\int_0^a \frac{(3a^2 - 2x^2}{\sqrt{(a^2 - x^2)}}dx$$

$$= \frac{\sqrt{2}}{a}\int_0^a \left[\frac{2(a^2 - x^2)}{\sqrt{(a^2 - x^2)}} + \frac{a^2}{\sqrt{(a^2 - x^2)}}\right]dx$$

$$= \frac{\sqrt{2}}{a}\int_0^a 2\sqrt{(a^2 - x^2)}dx + \frac{\sqrt{2}}{a}\int_0^a \frac{a^2}{\sqrt{(a^2 - x^2)}}dx$$

$$= \frac{2\sqrt{2}}{a}\int_0^a 2\sqrt{(a^2 - x^2)}\,dx + \frac{\sqrt{2}}{a}\int_0^a \frac{a^2}{\sqrt{(a^2 - x^2}}dx$$

$$= \frac{2\sqrt{2}}{a}\left[\frac{x\sqrt{(a^2 - x^2)}}{2} + \frac{a^2}{2}\sin^{-1}\frac{x}{a}\right]_0^a + \sqrt{2}.a\left[\sin^{-1}\frac{x}{a}\right]_0^a$$

$$= \frac{2\sqrt{2}}{a}\left[0 + \frac{a^2}{2}\cdot\frac{\pi}{2}\right] = \sqrt{2} \times a \times \frac{\pi}{2} = \pi a\sqrt{2}.$$

Example 37:

Find the length of the astroid $x^{2\ 3} + y^{2\ 3} = a^{2\ 3}$.

Solution:

The given astroid is $x^{2/3} + y^{2/3} = a^{2/3}$. ...(1)

The curve is symmetrical in all the four quadrants. For the arc of the curve in the first quadrant x varies from 0 to a. Differentiating (1), w.r.t. x, we get

$$\frac{2}{3}x^{-1/3} + \frac{2}{3}y^{-1/3}\frac{dy}{dx} = 0$$

so that $\frac{dy}{dx} = -\left(\frac{y}{x}\right)^{1/3}$

$\therefore$ the required whole length of the curve

=4 × length of the curve lying in the Ist quadrant

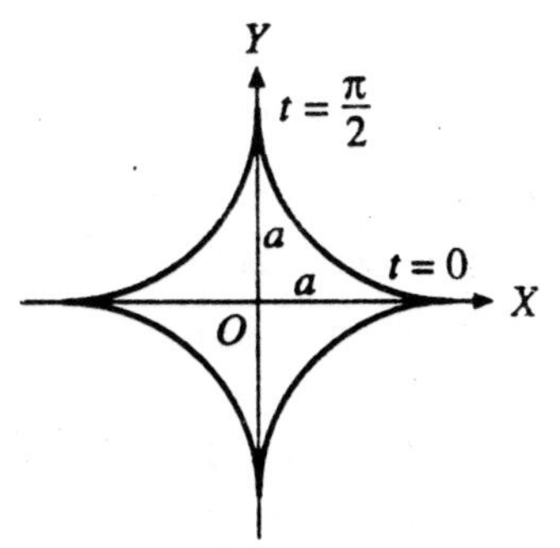

Fig. 2.6

$$= 4\int_0^a \sqrt{\left\{1+\left(\frac{dy}{dx}\right)^2\right\}}dx = 4\int_0^a \sqrt{\left(1+\frac{y^{2/3}}{x^{2/3}}\right)}dx$$

$$= 4\int_0^a \frac{\sqrt{(x^{2/3}+y^{2/3})}}{x^{1/3}}dx = 4\int_0^a \frac{\sqrt{(a^{2/3})}}{x^{1/3}}dx$$

$$= 4a^{1/3}\int_0^a x^{-1/3}dx = 4a^{1/3}\left[\frac{3}{2}x^{2/3}\right]_0^a = 6a.$$

Example 38:

Prove that the length of the curve $x^{2/3} + y^{2/3} = a^{2/3}$ measured from (0, a) to the point (x, y) is given by $s = 3/2(ax^2)^{1/3}$.

Solution:

Do your self.

The arc length of the astroid from x = 0 to x = x is obtained as

$$s = \int_0^x a^{1/3}x^{-1/3}dx = \left[a^{1/3}\frac{3}{2}x^{2/3}\right]_0^x = \frac{3}{2}(ax^2)^{1/3}.$$

Example 39:

Find the length of one quadrant of the curve $(x/a)^{2/3} + (y/b)^{2/3} = 1$.

Solution:

Do your self.

The parametric equations of the given curve are $x = a\cos^3 t$, $y = b\sin^3 t$.

Example 40:

Show that 8a is the length of an arch of the cycloid whose equations are $x = a(t - \sin t)$, $y = a(1 - \cos t)$.

Solution:

The given equation of the cycloid are $x = a(t - \sin t)$, $y = a(1 - \cos t)$. We have $dx/dt = a(1 - \cos t)$, and $dy/dt = a \sin t$.

$$\therefore \quad \frac{dy}{dx} = \frac{dy/dt}{dx/dt} = \frac{a\sin t}{a(1-\cos t)} = \frac{2\sin\frac{1}{2}t\cos\frac{1}{2}t}{2\sin^2\frac{1}{2}t} = \cot\frac{1}{2}t.$$

Now y = 0 when cos t = 1 *i.e.*, t = 0. At t = 0, x = 0, y = 0 and $dy/dx = \infty$. Thus the curve passes through the point (0, 0) and the tangent there is perpendicular to the x-axis.

Again y is maximum when cost = – 1 *i.e.*, t = π. When t = π, x = aπ. y = 2a, dy/dx = 0. Thus, at the point (aπ, 2a) the tangent to the curve is parallel to the x-axis.

Also in this curve y cannot be negative. Thus, an arch OBA of the given cycloid is as shown in the figure. It is symmetrical about the line BM which is the axis of the cycloid.

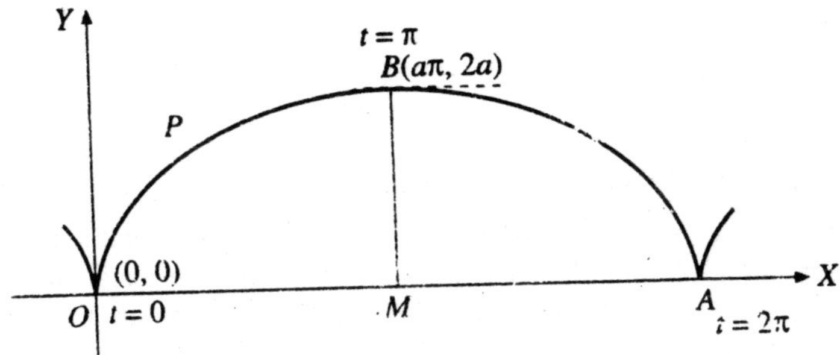

Fig. 2.7

We have $$\left(\frac{ds}{dt}\right)^2 = \left(\frac{dx}{dt}\right)^2 + \left(\frac{dy}{dt}\right)^2$$

$$= \{a\,(1 - \cos t)\}^2 + (a \sin t)^2$$

$$= a^2 \left\{\left(2 \sin^2 \frac{1}{2}t\right)^2 + \left(2 \sin \frac{1}{2}t \cos \frac{1}{2}t\right)^2\right\}$$

$$= 4a^2 \sin^2 \frac{1}{2}t \left(\sin^2 \frac{1}{2}t + \cos^2 \frac{1}{2}t\right) = 4a^2 \sin^2 \frac{1}{2}t. \quad \text{...(1)}$$

If s denotes the arc length of the cycloid measured from the cusp O to any point P towards the vertex B, then s increases as t increases. Therefore ds/dt will be taken with positive sign. So taking square root of both sides of (1),

we have $ds/dt = 2a \sin \frac{1}{2}t$,

or $ds = 2a \sin \frac{1}{2}t\, dt$.

At the cups O, t = 0

and at the vertex B, t = π.

Now the length of the arch OBA = 2 × length of the arc OB

$$= 2\int_0^{\pi} 2a \sin\frac{1}{2} t dt = 4a\left[-2\cos\frac{1}{2}t\right]_0^{\pi}$$

$$= 8a\left[\cos\frac{1}{2}t\right]_0^{\pi}$$

$$= -8a[0-1] = 8a.$$

Example 41:

Find the perimeter of the curve $x^2 + y^2 = a^2$.

Solution:

The equation of the curve is $x^2 + y^2 = a^2$. ...(1)

Here the curve is the standard equation of the circle with centre (0, 0) and radius a. Also it is symmetrical about both the axes. So the required perimeter will be four times the arc length lying in the first quadrant *i.e.*, between $x = 0$ to $x = a$.

Differentiating (1), w.r.t. x, we get $2x + 2y (dy/dx) = 0$

or $(dy/dx) = -(x/y) = -x/\sqrt{(a^2 - x^2)}$, from (1).

∴ the required perimeter

= 4 × {length of the arc in the first quadrant from (0, a) to (a, 0)}

$$= 4\int_{x=0}^{a}\sqrt{\left[1+\left(\frac{dy}{dx}\right)^2\right]}dx$$

$$= 4\int_0^a \sqrt{\left[1+\frac{x^2}{a^2-x^2}\right]}dx, \text{ putting for } \frac{dy}{dx}$$

$$= 4a\int_0^a \frac{dx}{\sqrt{(a^2-x^2)}} = 4a\left[\sin^{-1}\left(\frac{x}{a}\right)\right]_0^a$$

$$= 4a[\sin^{-1}(1) - \sin^{-1}(0)]$$

$$= 4a\left[\frac{1}{2}\pi - 0\right] = 2a\pi.$$

Example 42:

Find the length of the arc of the semi-cubical parabola $ay^2 = x^3$ *from the vertex to the point (a, a).*

Solution:

The given curve is $ay^2 = x^3$. ...(1)

It is symmetrical about the x-axis. We have to find the length of the arc from x = 0 to x = a in the first quadrant.

Differentiating (1) w.r.t. x, we get 2ay (dy/dx) = $3x^2$.

$$\therefore \quad \frac{dy}{dx} = \frac{3x^2}{2ay} = \frac{3x^2}{2a(x^3/a)^{1/2}}, \text{ substituting for y from (1)}$$

$$= \frac{3}{2}(x^{1/2} / a^{1/2}).$$

$\therefore$ required length

$$= \int_{x=0}^{a} \sqrt{\left[1 + \left(\frac{dy}{dx}\right)^2\right]} dx$$

$$= \int_0^a \sqrt{\left\{1 + \frac{9x}{4a}\right\}} dx = \frac{1}{2\sqrt{a}} \int_0^a (9x + 4a)^{1/2} dx$$

$$= \frac{1}{2\sqrt{a}} \left[\frac{2}{3} \cdot (9x + 4a)^{3/2} \cdot \frac{1}{9}\right]_0^a$$

$$= \frac{1}{27\sqrt{a}} [(13a)^{3/2} - (4a)^{3/2}]$$

$$= (a/27).[13\sqrt{(13)} - 8].$$

Example 43:

Show that the length of the arc of the curve $x^2 = a^2 (1 - e^{y/a})$ measured from the origin to the point (x, y) is a log {(a + x)/(a − x)} − x.

Solution:

The given equation of the curve is $x^2 = a^2 (1 - e^{y/a})$

or $$ey/a = 1 - \frac{x^2}{a^2}$$

or $$\frac{y}{a} = \log\left(1 - \frac{x^2}{a^2}\right)$$

or $$y = a \log\left(\frac{a^2 - x^2}{a^2}\right)$$

or $$y = a \log (a^2 - x^2) - a \log a^2 \qquad \ldots(1)$$

Differentiating (1) w.r.t. x, we get

$$\frac{dy}{dx} = a \cdot \frac{-2x}{a^2 - x^2}.$$

$\therefore$ required arc length

$$= \int_0^x \left[1 + \left(\frac{dy}{dx}\right)^2\right]^{1/2} dx = \int_0^x \left[1 + \frac{4a^2x^2}{(a^2 - x^2)^2}\right]^{1/2} dx$$

$$= \int_0^x \left(\frac{a^2 + x^2}{a^2 - x^2}\right) dx = \int_0^x \left(-1 + \frac{2a^2}{a^2 - x^2}\right) dx$$

$$= \left[-x + 2a^2 \cdot \frac{1}{2a} \log \frac{a + x}{a - x}\right]_0^x = a \log \frac{a + x}{a - x} - x.$$

Example 44:

Rectify the curve or find the length of an arch of the curve

$x = a\ (t + \sin t),\ y = a\ (1 - \cos t).$

Solution:

Differentiating the given parametric equation of the cycloid w.r.t. t, we have $dx/dt = a\ (1 + \cos t)$, and $dy/dt = a \sin t$.

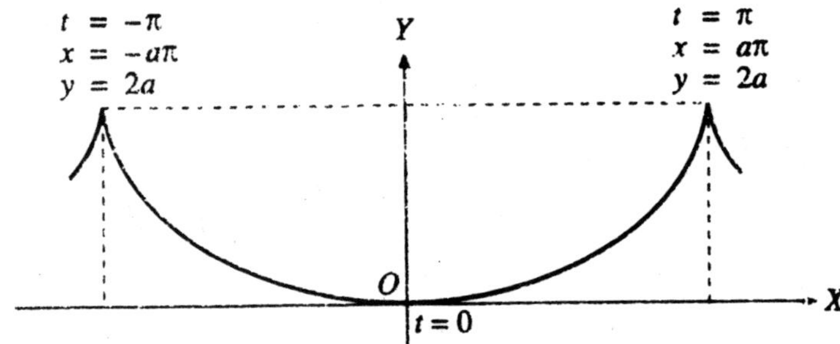

Fig. 2.8

If we measure the arc length s in the direction of t increasing we have

$$\frac{ds}{dt} = \sqrt{\left\{\left(\frac{dx}{dt}\right)^2 + \left(\frac{dy}{dt}\right)^2\right\}}$$

$$= \sqrt{\{a^2\ (1 + \cos t)^2 + a^2 \sin^2 t\}}$$

$$= a\ \{1 + \cos^2 t + 2 \cos t + \sin^2 t\}^{1/2}$$

$$= a\sqrt{2}\ (1 + \cos t)^{1/2} = a\sqrt{2}\left(2\cos^2 \frac{1}{2}t\right)^{1/2}$$

$$= 2a \cos \frac{1}{2} t.$$

$$\therefore \quad ds = 2a \cos \frac{1}{2} t \, dt.$$

For an arch of the given cycloid lying between two successive cusps t varies from $-\pi$ to π. Also this arch is symmetrical about the y-axis and we have t = 0 at the vertex O.

$\therefore$ the required whole length of the arch

$$= 2 \times \text{length of the arc from } t = 0 \text{ to } t = \pi$$

$$= 2\int_0^{\pi} 2a \cos \frac{1}{2} t dt = 4a \int_0^{\pi} \cos \frac{1}{2} t dt$$

$$= 4a \left[2 \sin \frac{1}{2} t \right]_0^{\pi}$$

$$= 4a \cdot [2 - 0] = 8a.$$

Example. 45:

Prove that the length of an arc of the cycloid $x = a (t + \sin t)$, $y = a (1 - \cos t)$ form the vertex to the point (x, y) is $\sqrt{(8ay)}$.

Solution:

Let s denote the arc length of the cycloid measured from the vertex to any point P (*i.e.*, form t = 0 to t = t).

Then proceeding as in Ex. 20, the required arc length

$$s = \int_0^t \sqrt{\left\{ \left(\frac{dx}{dt} \right)^2 + \left(\frac{dy}{dt} \right)^2 \right\}} dt = 2a \int_0^t \cos \frac{1}{2} t dt$$

$$= 2a \left[2 \sin \frac{1}{2} t \right]_0^t = 4a \sin \frac{1}{2} t = \sqrt{\left(8a^2 \cdot 2 \sin \frac{1}{2} t \right)}$$

$$= \sqrt{\{8a \times a (1 - \cos t)\}} = \sqrt{(8ay)},$$

$$[\because y = a (1 - \cos t)].$$

Example 46:

Find the perimeter of the cardioid $r = a (1 - \cos \theta)$.

Solution:

The given curve is $r = a (1 - \cos \theta)$

It is symmetrical about the initial line.

We have $r = 0$ when $\cos \theta = 1$

i.e., $\theta = 0$.

Also r is maximum when $\cos\theta = -1$

i.e., $\theta = \pi$ and then $r = 2a$.

As θ increases from 0 to π, r increases from 0 to 2a. So the curve is as shown in the figure.

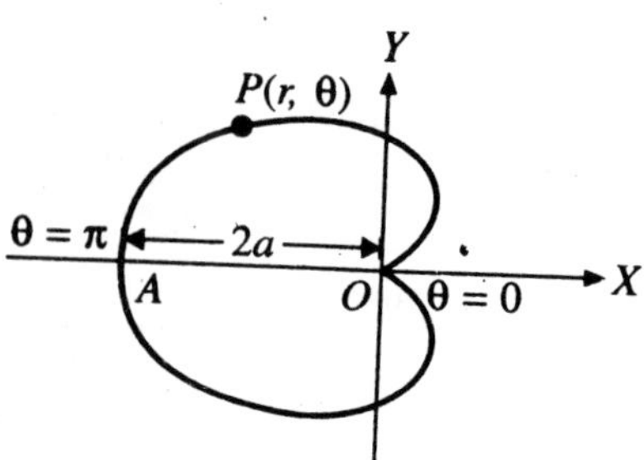

Fig. 2.9

By symmetry, the perimeter of the cardioid

=2 × the arc length of the upper half of the cardioid.

Now differentiating (1) w.r.t. θ, we have $dr/d\theta = a \sin\theta$.

We have $$\left(\frac{ds}{d\theta}\right)^2 = r^2 + \left(\frac{dr}{d\theta}\right)^2 = a^2(1-\cos\theta)^2 + a^2\sin^2\theta$$

$$= a^2\left(2\sin^2\frac{1}{2}\theta\right)^2 + a^2\left(2\sin\frac{1}{2}\theta\cos\frac{1}{2}\theta\right)^2$$

$$= 4a^2\sin^2\frac{1}{2}\theta\left(\sin^2\frac{1}{2}\theta + \cos^2\frac{1}{2}\theta\right)$$

$$= 4a^2\sin^2\frac{1}{2}\theta.$$

If s denotes the arc length of the cardioid measured from the cusp O (*i.e.*, the point $\theta = 0$) to any point P (r, θ) in the direction of θ increasing, then s increases as θ increases. Therefore $ds/d\theta$ will be positive.

Hence from (2), we have $ds/d\theta = 2a\sin 1/2\theta$.

or $ds = 2a\sin 1/2\theta\, d\theta$

At the cusp O, $\theta = 0$

and at vertex A, $\theta = \pi$.

∴ the length of the arc OPA $= \int_0^{\pi} 2a\sin\frac{1}{2}\theta d\theta$

$$= 4a\left[-\cos\frac{\theta}{2}\right]_0^{\pi} = -4a\left[\cos\frac{\theta}{2}\right]_0^{\pi} = -4a(0-1) = 4a.$$

∴ the perimeter of the cardioid = 2 × 4a = 8a.

Example 47:

Find the entire length of the cardioid $r = a(1 + \cos\theta)$.

Solution:

The given curve is r = a (1 + cos θ). ...(1)

It is symmetrical about the initial line and for the portion of the curve lying above the initial line θ varies from θ = 0 to θ = π.

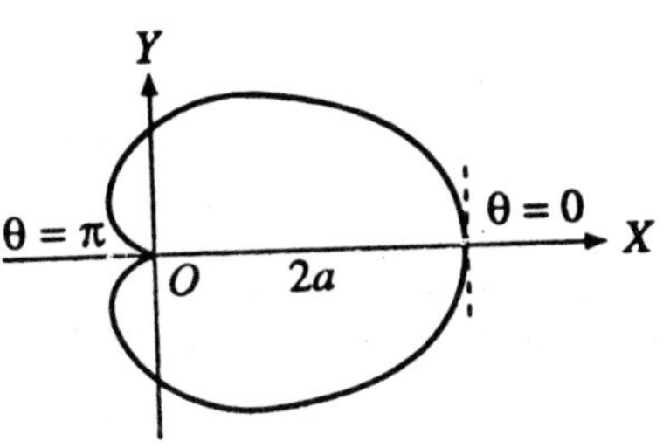

Fig. 2.10

Now differentiating (1) w.r.t. θ, we have dr/dθ = – a sin θ.

If s denotes the arc length of the cardioid measured from the vertex (*i.e.*, the point θ = 0) to any point P (r, θ) in the direction of θ increasing, we have

$$\frac{ds}{d\theta} = \sqrt{\left\{r^2 + \left(\frac{dr}{d\theta}\right)^2\right\}}$$

$$= \sqrt{\{a^2 (1 + \cos\theta)^2 + (-a\sin\theta)^2\}}$$

$$= \sqrt{\left\{a^2\left(2\cos^2\frac{1}{2}\theta\right)^2 + a^2\cdot\left(2\sin\frac{1}{2}\theta\cos\frac{1}{2}\theta\right)^2\right\}}$$

$$= 2a\sqrt{\left\{\cos^4\frac{1}{2}\theta + \sin^2\frac{1}{2}\theta\cos^2\frac{1}{2}\theta\right\}}$$

$$= 2a\sqrt{\left\{\cos^2\frac{1}{2}\theta\left(\cos^2\frac{1}{2}\theta + \sin^2\frac{1}{2}\theta\right)\right\}}$$

$$= 2a\sqrt{\left(\cos^2\frac{1}{2}\theta\right)} = 2a\cos\frac{1}{2}\theta.$$

$$\therefore \qquad ds = 2a\cos\frac{1}{2}\theta\, d\pi. \qquad ...(2)$$

Let s_1 denote the arc length of the upper half of the cardioid (*i.e.*, from θ = 0 to θ = π). Then

$$\int_0^{s_1} ds = \int_0^{\pi} 2a\cos\frac{1}{2}\theta d\theta = 2a\left[2\sin\frac{1}{2}\theta\right]_0^{\pi}$$

or
$$s_2 = 4a\left[\sin\frac{1}{2}\pi - \sin 0\right] = 4a\ (1 - 0) = 4a.$$

∴ by symmetry, the whole length of the cardioid = 2 × the arc length of the upper half of the cardioid = 2 · 4a = 8a.

Example 48:

Find the perimeter of the curve $r = a(1 + \cos\theta)$ and show that arc of thee upper half is bisected by $\theta = \pi/3$.

Solution:

Length s_1 of the upper half of the cardioid $r = a(1 + \cos\theta)$ is 4a.

$\therefore$ the perimeter of the cardioid $= 2 \times 4a = 8a$.

Again, if s_2 denotes the arc length of the cardioid from the point $\theta = 0$ to the pint $\theta = \pi/3$, then

$$s^2 = \int_0^{\pi/3} 2a\cos\frac{1}{2}\theta d\theta,$$

$$= 2a\left[2\sin\frac{1}{2}\theta\right]_0^{\pi/3} = 4a\left[\sin\frac{\pi}{6} - \sin 0\right] = 4a \times \frac{1}{2}$$

$$= \frac{1}{2}(s_1)$$

= half the arc length of the upper half of the cardioid.

Example 49:

Prove that the line $4r\cos\theta = 3a$ divides the cardioid $r = a(1 + \cos\theta)$ into two parts such that lengths of the arc on either side of the line are equal.

Solution:

The given equation of the cardioid and the line are

$r = a(1 + \cos\theta)$ and $4r\cos\theta = 3a$.

Eliminating r, we have $a(1 + \cos\theta) = 3a/(4\cos\theta)$

or $\quad 4\cos\theta + 4\cos^2\theta = 3$

or $\quad 4\cos^2\theta + 4\cos\theta - 3 = 0$

or $\quad (2\cos\theta - 1)(2\cos\theta + 3) = 0$

i.e., $\quad \cos\theta = \dfrac{1}{2}$ or $\cos\theta = -\dfrac{3}{2}$.

But the values of $\cos\theta$ cannot be numerically greater than 1.

Therefore $\cos\theta = -3/2$ is inadmissible.

So we have $\cos\theta = 1/2$ *i.e.,* $\theta = \pi/3$.

Hence the vectorial angle of the point of intersection of the cardioid with the given line is $\pi/3$. The cartesian equation of the given line is $x = 3a/4$ *i.e.,* the line is perpendicular to the x-axis. So now we have to prove that the are of the upper half of the cardioid is bisected by $\theta = \pi/3$.

Example 50:

Show that the arc of the upper half of the curve $r = a(1 - \cos\theta)$ is bisected by $\theta = 2\pi/3$.

Solution:

As proved in Ex. 31, the arc length of the upper half of the cardioid $r = a(1 - \cos\theta)$ is 4a.

Also, the arc length of the cardioid from the point $\theta = 0$ to the point $\theta = 2\pi/3$

$$= 2a\int_0^{2\pi/3} \sin\frac{1}{2}\theta d\theta,$$

$$= 2a\left[-2\cos\frac{1}{2}\theta\right]_0^{2\pi/3} = -4a\left[\cos\frac{\pi}{3} - \cos 0\right] = -4a\left[\frac{1}{2} - 1\right]$$

$= 2a = 1/2\ (4a) =$ half the arc length of the upper half of the cardioid.

Hence the arc of the upper half of the curve $r = a(1 - \cos\theta)$ is bisected by $\theta = 2\pi/3$.

Example 51:

Find the length of the cardioid $r = a(1 - \cos\theta)$ lying outside the circle $r = a\cos\theta$.

Solution:

The given cardioid is $r = a(1 - \cos\theta)$. ...(1)

It meets the circle $r = a\cos\theta$. ...(2)

Eliminating r between (1) and (2), we have

$$a(1 - \cos\theta) = a\cos\theta \text{ or } 1 - \cos\theta = \cos\theta$$

or $$2\cos\theta = 1$$

or $$\cos\theta = \frac{1}{2} \text{ i.e., } \theta = \frac{1}{3}\pi$$

Hence the vectorial angle of the point of intersection P of the cardioid $r = a(1 - \cos\theta)$ with the circle $r = a\cos\theta$ is $\pi/3$.

So for the portion of the cardioid $r = a(1 - \cos\theta)$ lying outside the circle $r = a\cos\theta$ (above the initial line) θ varies from $\theta = \pi/3$ to $\theta = \pi$.

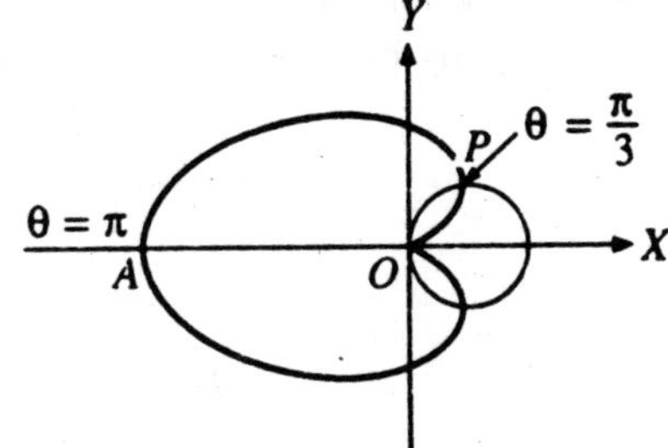

Fig. 2.11

Also differentiating (1), we get $dr/d\theta = -a\sin\theta$.

By symmetry, the required length of the cardioid = 2 × the arc length from $\theta = \pi/3$ to $\theta = \pi$ of the upper half of the cardioid

$$= 2\int_{\theta=\pi/3}^{\pi}\sqrt{\left\{r^2+\left(\frac{dr}{d\theta}\right)^2\right\}}d\theta$$

$$= 2\int_{\pi/3}^{\pi}\sqrt{\left\{a^2(1-\cos\theta)^2+(-a\sin\theta)^2\right\}}d\theta$$

$$= 2a\int_{\pi/3}^{\pi}\sqrt{\left\{\left(2\sin^2\frac{1}{2}\theta\right)^2+4\sin^2\frac{1}{2}\theta\cos^2\frac{1}{2}\theta\right\}}d\theta$$

$$= 4a\int_{\pi/3}^{\pi}\sqrt{\left\{\sin^2\frac{1}{2}\theta\left(\sin^2\frac{1}{2}\theta+\cos^2\frac{1}{2}\theta\right)\right\}}d\theta$$

$$= 4a\int_{\pi/3}^{\pi}\sin\frac{1}{2}\theta d\theta = 4a\left[-2\cos\frac{1}{2}\theta\right]_{\pi/3}^{\pi}$$

$$= -8a\left[\cos\frac{1}{2}\pi-\cos\frac{1}{6}\pi\right] = -8a\left[0-\frac{1}{2}\sqrt{3}\right] = 4a\sqrt{3}.$$

Example 52:

Find the length of the arc of the equiangular spiral $r = a\,e^{\theta\cot a}$, taking $s = 0$ when $\theta = 0$.

Solution:

The given equiangular spiral is $r = ae^{\theta\cot\alpha}$. ...(1)

Differentiating (1) w.r.t. θ,

we get $dr/d\theta = ae^{\theta\cot\alpha}.\cot\alpha = r\cot\alpha$.

If s denotes the arc length of the equiangular spiral measured from $\theta = 0$ to any point P (r, θ) in the direction of θ increasing, we have

$$\frac{ds}{d\theta} = \sqrt{\left\{r^2+\left(\frac{dr}{d\theta}\right)^2\right\}} = \sqrt{(r^2+r^2\cot^2\alpha)} = r\,\text{cosec}\,\alpha$$

$$= ae^{\theta\cot\alpha}.\ \text{cosec}\,\alpha.$$

or $$ds = ae^{\theta\cot\alpha}\times\text{cosec}\,\alpha\,d\theta.$$

$\therefore$ Integrating, $\int_0^s ds = a\,\text{cosec}\,\alpha\int_0^\theta e^{\theta\cot\alpha}d\theta$

or $$s = a \operatorname{cosec} \alpha . \frac{1}{\cot \alpha} \cdot \left[e^{\theta \cot \alpha} \right]_0^{\theta}$$
$$= a \sec \alpha \; [e^{\theta \cot \alpha} - e^0] = a \sec \alpha \; [e^{\theta \cot \alpha} - t].$$

Example 53:

Find the length of the arc of the equiangular spiral $r = ae^{\theta \cot \alpha}$ between the points for which radii vectors are r_1 and r_2.

Solution:

The given curve is $r = ae^{\theta \cot \alpha}$. ...(1)

Differentiating (1) w.r.t. θ, we get

$$dr/d\theta = ae^{\theta \cot \alpha} . \cot \alpha = r \cot \alpha, \quad \text{from (1).}$$

$$\therefore \quad d\theta/dr = 1/(r \cot \alpha) \; \textit{i.e.,} \; (r \, d\theta/dr) = \tan \alpha. \qquad ...(2)$$

If s denotes the arc length of the given curve measured in the direction of r increasing, we have

$$\frac{ds}{dr} = \sqrt{\left\{1 + r^2 \left(\frac{d\theta}{dr}\right)^2\right\}}$$
$$= \sqrt{(1 + \tan^2 \alpha)} = \sqrt{(\sec^2 \alpha)} = \sec \alpha,$$

or $$ds = \sec \alpha \, dr.$$

Let s_1 denote the required arc length *i.e.,* from $r = r_1$ to $r = r_2$.

Then $$\int_0^{s_1} ds = \int_{r_1}^{r_2} \sec \alpha dr = (\sec a) \left[r \right]_{r_1}^{r_2}$$

or $$s_1 = (\sec \alpha) \, (r_2 - r_1).$$

Example 54:

Find the length of any arc of the cissoid $r = a \, (\sin^2 \theta / \cos \theta)$.

Solution:

The given curve is $r = a \, (\sin^2 \theta / \cos \theta)$

or $$r = a \tan \theta \sin \theta. \qquad ...(1)$$

Differentiating (1) w.r.t. θ, we have

$$dr/d\theta = a \, [\tan \theta \cos \theta + \sec^2 \theta \sin \theta]$$
$$= a \, [\sin \theta + \sec^2 \theta \sin \theta] = a \sin \theta \; [1 + \sec^2 \theta].$$

We have $$(ds/d\theta)^2 = r^2 + (dr/d\theta)^2$$
$$= a^2 \tan^2 \theta \sin^2 \theta + a^2 \sin^2 \theta \; (1 + \sec^2 \theta)^2$$

$$= a^2 \sin^2\theta \,\{\tan^2\theta + (1 + \sec^2\theta)^2\}$$

$$= a^2 \sin^2\theta \,[\tan^2\theta + 1 + \sec^4\theta + 2\sec^2\theta]$$

$$= a^2 \sin^2\theta \,[\sec^2\theta + \sec^4\theta + 2\sec^2\theta]$$

$$= a^2 \sin^2\theta . \sec^2\theta \,[3 + \sec^2\theta]$$

$$= a^2 \tan^2\theta \,[3 + \sec^2\theta]. \qquad ...(2)$$

If s denotes the arc length of the cissoid measured from the point $\theta = \theta_1$ in the direction of θ increasing, then $ds = a \tan\theta \sqrt{(3 + \sec^2\theta)}\, d\theta$, on taking square root of (2) and keeping the +ive sign.

Let s_1 denote the required arc length from $\theta = \theta_1$ to $\theta = \theta_2$. Then

$$\int_0^{s_1} ds = \int_{\theta=\theta_1}^{\theta_2} a \tan\theta \sqrt{(3+\sec^2\theta)}\,d\theta$$

$$= a\int_{\theta_1}^{\theta_2} \frac{\sin\theta}{\cos\theta}\sqrt{\left(3+\frac{1}{\cos^2\theta}\right)}d\theta$$

or

$$s_1 = a\int_{\theta_1}^{\theta_2} \frac{\sin\theta}{\cos^2\theta}\sqrt{\left(1+3\cos^2\theta\right)}d\theta \qquad ...(3)$$

Let us first evaluate the indefinite integral

$$I = \int \frac{\sin\theta}{\cos^2\theta}\sqrt{\{1+3\cos^2\theta\}}d\theta.$$

Put $\cos\theta = t$ so that $-\sin\theta\, d\theta = dt$.

Then
$$I = -\int \frac{\sqrt{(1+3t^2)}}{t^2}dt = -\int \frac{(1+3t^2)}{t^2}dt$$

$$= -\int \frac{dt}{t^2\sqrt{(1+3t^2)}} - \int \frac{3dt}{\sqrt{(1+3t^2)}}$$

To evaluate the first integral, put $t = 1/z$ so that $dt = -(1/z^2)\,dz$.

Then
$$-\int \frac{dt}{t^2\sqrt{(1+3t^2)}} = \int \frac{dz}{\sqrt{(1+3/z^2)}}$$

$$= \int \frac{2z\,dz}{t^2\sqrt{(z^2+3)}}$$

$$= \sqrt{(z^2 + 3)}, \text{ by power formula}$$

$$= \sqrt{\left(\frac{1}{t^2}+3\right)} = \sqrt{\left(\frac{1}{\cos^2}+3\right)} = \sqrt{(\sec^2\theta + 3)}.$$

Also the second integral

$$= -\int \frac{3dt}{\sqrt{(1+3t^2)}} = -\sqrt{3}\int \frac{dt}{\sqrt{\left(\frac{1}{3}+t^2\right)}}$$

$$= -\sqrt{3}\log\left\{t+\sqrt{\left(t^2+\frac{1}{3}\right)}\right\}$$

$$= -\sqrt{3}\log\left\{\cos\theta+\sqrt{\left(\cos^2\theta+\frac{1}{3}\right)}\right\}.$$

$\therefore$ $$I = \sqrt{(\sec^2\theta+3)} - \sqrt{3}\log\left\{\cos\theta+\sqrt{\left(\cos^2\theta+\frac{1}{3}\right)}\right\}.$$

Hence from (3), we get the required arc length

$$s_1 = \left[a\sqrt{\left(\sec^2\theta+3\right)} - \sqrt{3}.a\log\left\{\cos\theta+\sqrt{\left(\cos^2+\frac{1}{3}\right)}\right\}\right]_{\theta_1}^{\theta_2}$$

$$= f(\theta_2) - f(\theta_1),$$

where $f(\theta) = a\sqrt{(\sec^2\theta + 3)} - a\sqrt{3}\log\left\{\cos\theta+\sqrt{\left(\cos^2\theta+\frac{1}{3}\right)}\right\}$.

Example 55:

Find the length of the curve $r^{l\,3} = 8\cos(\theta/3)$.

Solution:

The given curve is $r^{1/3} = 8\cos(\theta/3)$

or $$r = 512\cos^3(\theta/3). \quad ...(1)$$

The curve (1) is symmetrical about the initial line.

We have $r = 0$ when $\cos(\theta/3) = 0$

i.e., $\theta/3 = \pm\pi/2$

i.e., $\theta = -3\pi/2$

and $\theta = 3\pi/2$.

The entire length of the curve = 2. (length of the curve from $\theta = 0$ to $\theta = 3\pi/2$).

From (1),

$$\frac{dr}{d\theta} = 512.3\left(\cos^2\frac{\theta}{3}\right)\left(-\sin\frac{\theta}{3}\right)\cdot\frac{1}{3}$$

$$= -512\cos^2\frac{\theta}{3}\sin\frac{\theta}{3}.$$

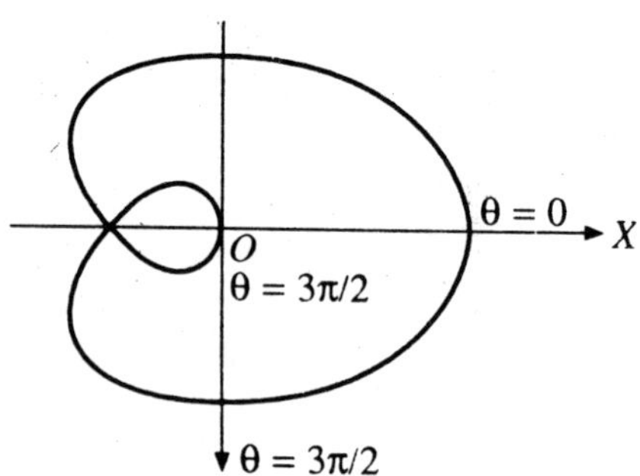

Fig. 2.12

Now $$\left(\frac{ds}{d\theta}\right)^2 = r^2 + \left(\frac{dr}{d\theta}\right)^2$$

$$= (512)^2 \cos^6 \frac{\theta}{3} + (512)^2 \cos^4 \frac{\theta}{3} \sin^2 \frac{\theta}{3}.$$

$$= (512)^2 \cos^4\theta \frac{\theta}{3}\left(\cos^2 \frac{\theta}{3} + \sin^2 \frac{\theta}{3}\right)$$

$$= (512)^2 \cos^4 \frac{\theta}{3}$$

$\therefore$ $$\frac{ds}{d\theta} = 512 \cos^2 \frac{\theta}{3}$$

or $$ds = 512 \cos^2 \frac{\theta}{3} d\theta.$$

$\therefore$ the required perimeter of the curve

$$= 2\int_0^{3\pi/2} 512 \cos^2 \frac{\theta}{3} d\theta$$

$$= 6.512\int_0^{\pi/2} \cos^2 t dt,$$

putting $\frac{\theta}{3} = t$ so that $d\theta = 3dt$

$$= 3072 \times \frac{1}{2}\cdot\frac{\pi}{2} = 768\pi.$$

Example 56:

Prove that the perimeter of the limacon $r = a + b \cos\theta$, if b/a be small, is approximately $2\pi a\left(1 + \frac{1}{4}b^2 / a^2\right)$.

Solution:

The given curve is $r = a + b \cos \theta$, $(a > b)$. ...(1)

Note that b/a is given to be small so we must have $b < a$. The curve (1) is symmetrical about the initial line and for the portion of the curve lying above the initial line θ varies from $\theta = 0$ to $\theta = \pi$.

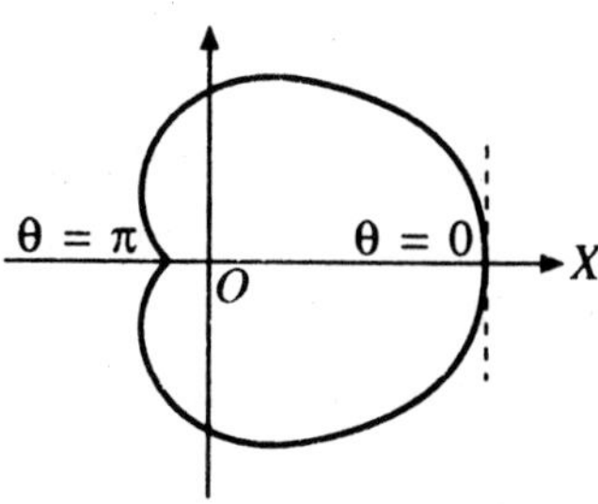

Fig. 2.13

By symmetry, the perimeter of the limacon = 2 × the arc length of the upper half of the limacon.

Now differentiating (1) w.r.t. θ, we have $dr/d\theta = - b \sin \theta$.

We have
$$\left(\frac{ds}{d\theta}\right)^2 = r^2 + \left(\frac{dr}{d\theta}\right)^2$$
$$= (a + b\cos\theta)^2 + (-b\sin\theta)^2$$
$$= a^2 + b^2\cos^2\theta + 2ab\cos\theta + b^2\sin^2\theta$$
$$= a^2 + b^2 + 2ab\cos\theta.$$

If we measure the arc length s in the direction of θ increasing, we have $ds/d\theta = \sqrt{(a^2 + b^2 + 2ab\cos\theta)}$

or
$$ds = \sqrt{(a^2 + b^2 + 2ab\cos\theta)}\, d\theta.$$

The arc length of the upper half of the limacon

$$= \int_0^\pi \sqrt{(a^2 + b^2 + 2ab\cos\theta)}\,d\theta = a\int_0^\pi \left(1 + \frac{2b}{a}\cos\theta + \frac{b^2}{a^2}\right)^{1/2} d\theta$$

$$= a\int_0^\pi \left[1 + \frac{b}{a}\cos\theta + \frac{1}{2}\cdot\frac{b^2}{a^2} + \frac{\frac{1}{2}\left(\frac{1}{2}-1\right)}{2!}\left(4\frac{b^2}{a^2}\cos^2\theta\right)\right]d\theta,$$

expanding by binomial theorem and neglecting powers of b/a higher than two because b/a is small

$$= a\int_0^\pi \left[1 + \frac{b}{a}\cos\theta + \frac{1}{2}\frac{b^2}{a^2}\left(1 - \cos^2\theta\right)\right]d\theta$$

$$= a\int_0^\pi \left[1 + \frac{b}{a}\cos\theta + \frac{1}{2}\frac{b^2}{a^2}\sin^2\theta\right]d\theta$$

$$= a\left[\left\{\theta + \frac{b}{a}\sin\theta\right\}_0^{\theta} + \frac{1}{2}\frac{b^2}{a^2}2\int_0^{\pi/2}\sin^2\theta d\theta\right]$$

$$= a\left[\pi + \frac{1}{2}\frac{b^2}{a^2}\cdot 2\cdot\frac{1}{2}\cdot\frac{\pi}{2}\right] = a\pi\left[1 + \frac{b^2}{4a^2}\right].$$

$\therefore$ the perimeter of the limacon

$= 2 \times a\pi\ [1 + (b^2/4a^2)] = 2a\pi\ [1 + (b^2/4a^2)]$.

Example 57:

Show that the whole length of the limacon $r = a + b\cos\theta$, $(a > b)$ *is equal to that of an ellipse whose semi-axes are equal in length to the maximum and minimum radii vectors of the limacon.*

Solution:

The whole length of the limacon

$$= 2\int_0^{\pi}\sqrt{\left(a^2 + b^2 + 2ab\cos\theta\right)}d\theta, \text{ (prove it here).} \qquad ...(1)$$

Also the maximum and minimum radii vectors of the limacon are given by $\cos\theta = 1$ and $\cos\theta = -1$ and the they are respectively $a + b$ and $a - b$.

Now, the parametric equations of the ellipse with major axis as $(a + b)$ and minor axis as $(a - b)$ may be taken as

$$x = (a + b)\cos\phi,\ y = (a - b)\cos\phi. \qquad ...(2)$$

Differentiating (2) w.r.t. ϕ, we have

$$dx/d\phi = -(a + b)\sin\phi,\ dy/d\phi = (a - b)\cos\phi.$$

Now the ellipse (2) is symmetrical in all the four quadrants and for the portion of the ellipse lying in the first quadrant θ varies from

$$\phi = 0 \text{ to } \phi = 1/2\pi.$$

By symmetry, the perimeter (whole length) of the ellipse = 4 × the arc length of the ellipse lying in the first quadrant

$$= 4\int_0^{\pi/2}\sqrt{\left\{\left(\frac{dx}{d\phi}\right)^2 + \left(\frac{dy}{d\phi}\right)^2\right\}}d\phi$$

$$= 4\int_0^{\pi/2}\sqrt{\left[\{-(a+b)\sin\phi\}^2 + \{(a-b)\cos\phi\}^2\right]}d\phi$$

$$= 4\int_0^{\pi/2} \sqrt{\{a^2 \sin^2 \phi + b^2 \sin^2 \phi + 2ab \sin^2 \phi + a^2 \cos^2 \phi + b^2 \cos^2 \phi - 2ab \cos^2 \phi\}}\, d\phi$$

$$= 4\int_0^{\pi/2} \sqrt{\{a^2 + b^2 - 2ab(\cos^2 \phi - \sin^2 \phi)\}}\, d\phi$$

$$= 4\int_0^{\pi/2} \sqrt{\{a^2 + b^2 - 2ab \cos 2\phi\}}\, d\phi.$$

Now put $2\phi = t$ so that $2d\phi = dt$.

Also when $\phi = 0$, $t = 0$

and when $\phi = \pi/2$, $t = \pi$.

Then the whole length of the ellipse

$$= 4\int_0^{\pi} \sqrt{\{a^2 + b^2 - 2ab \cos t\}}\, \frac{1}{2} dt$$

$$= 2\int_0^{\pi} \sqrt{\{a^2 + b^2 - 2ab \cos t\}}\, dt$$

$$= 2\int_0^{\pi} \sqrt{\{a^2 + b^2 - 2ab \cos(\pi - t)\}}\, dt,$$

$$\left[\because \int_0^a f(x)dx = \int_0^a f(a - x)dx\right]$$

$$= 2\int_0^{\pi} \sqrt{(a^2 + b^2 + 2ab \cos t)}\, dt$$

$$= 2\int_0^{\pi} \sqrt{(a^2 + b^2 + 2ab \cos \theta)}\, d\theta,$$

$$\left[\because \int_0^a f(x)dx = \int_0^a f(t)dt\right]$$

= the whole length of the limacon, [from (1)].

Example 58:

Show that the length of a loop of the curve

$3x^2y - y^3 = (x^2 + y^2)^3$ *is* $2\int_0^1 \frac{dr}{\sqrt{(1 - r^6)}}$.

Solution:

Changing to polar form by putting $x = r \cos \theta$, $y = r \sin \theta$, the given equation of the curve becomes

$$3r^2 \cos^2 \theta \times r \sin \theta - r^3 \sin \theta = (r^2)^3$$

or $r^3 (3 \sin\theta \cos^2\theta - \sin^3\theta) = r^6$

or $3\sin\theta\,(1 - \sin^2\theta) - \sin^3\theta = r^3$

or $3\sin\theta - 3\sin^3\theta - \sin^3\theta = r^3$ or $r^3 = 3\sin\theta - 4\sin^3\theta$

or $r^3 = \sin 3\theta.$ ...(1)

Differentiating (1) w.r.t. θ, we have

$$3r^2 (dr/d\theta) = 3\cos 3\theta \text{ or } (dr/d\theta) = (\cos 3\theta)/r^2.$$

$$\therefore \quad r\frac{d\theta}{dr} = r.\frac{r^2}{\cos 3\theta} = \frac{r^3}{\cos 3\theta}$$

$$= \frac{\sin 3\theta}{\cos 3\theta} = \tan 3\theta, \qquad \text{from (1).}$$

From (1), r = 0 when sin 3θ = 0 *i.e.*, when 3θ = 0, π *i.e.*, when θ = 0, π/3. Thus, two consecutive values of θ for which r = 0 are 0 and θ/3. Therefore one loop of the curve lies between θ = 0 and π/3. Also r is maximum when sin 3θ = 1 *i.e.*, 3θ = π/2 *i.e.*, θ = π/6. Therefore half of the loop extends from θ = 0 to θ = π/6.

When θ = 0, r = 0

and when θ = π/6, r = 1.

∴ the required length of a loop

$$= 2\int_0^1 \sqrt{\left\{1 + r^2\left(\frac{d\theta}{dr}\right)^2\right\}}dr, \qquad \text{(by symmetry)}$$

$$= 2\int_0^1 \sqrt{\{1+\tan^2 3\theta\}}dr = 2\int_0^1 \sec 3\theta\, dr = 2\int_0^1 \frac{dr}{\cos 3\theta}.$$

$$= 2\int_0^1 \frac{dr}{\sqrt{(1-\sin^2 3\theta)}} = 2\int_0^1 \frac{dr}{\sqrt{(1-r^6)}}, \qquad \text{from (1).}$$

Example 59:

If s be the length of the curve r = a tanh 1/2θ between the origin and θ = 2π, and Δ be the area under the curve between the same two points, prove that Δ = a (s – aπ).

Solution:

The given curve is r = a tanh 1/2θ. ...(1)

Differentiating (1) w.r.t. θ,

we get dr/dθ = a × 1/2 sec h^2 1/2θ.

We have $$\left(\frac{ds}{d\theta}\right)^2 = r^2 + \left(\frac{dr}{d\theta}\right)^2$$

$$= a^2 \tanh^2 \frac{1}{2}\theta + \frac{a^2}{4} \operatorname{sech}^4 \frac{1}{2}\theta$$

$$= \frac{1}{4}a^2 \left[4 \tanh^2 \frac{1}{2}\theta + \sec h^4 \frac{1}{2}\theta\right]$$

$$= \frac{1}{4}a^2 \left[4\left(1 - \sec h^2 \frac{1}{2}\theta\right) + \sec h^4 \frac{1}{2}\theta\right]$$

$$= \frac{1}{4}a^2 \left[2 - \sec h^2 \frac{1}{2}\theta\right]^2. \qquad ...(2)$$

If we measure the arc length s in the direction of θ increasing, we have

$$ds/d\theta = \frac{1}{2}a \left(2 - \sec h^2 \frac{1}{2}\theta\right),$$

retaining +ive sign while taking the square root of (2)

or $$ds = \frac{1}{2}a \left(2 - \sec h^2 \frac{1}{2}\theta\right)d\theta.$$

Now at the origin $r = 0$

and putting $r = 0$ in (1), we get $\theta = 0$.

∴ the arc length of the given curve between the origin ($\theta = 0$) and $\theta = 2\pi$ is given by

$$s = \frac{1}{2}a\int_0^{2\pi}\left(2 - \sec h^2 \frac{1}{2}\theta\right)d\theta$$

$$= \frac{1}{4}a\int_0^{2\pi} 2d\theta - \frac{1}{2}a\int_0^{2\pi} \sec h^2 \frac{1}{2}\theta d\theta$$

$$= \frac{1}{2}a \cdot 2[\theta]_0^{2\pi} - \frac{1}{2}a\left[2 \tanh \frac{1}{2}\theta\right]_0^{2\pi}$$

$$= 2a\pi - a \tan h\, \pi. \qquad ...(3)$$

Also the area between the radii vectors $\theta = 0$, $\theta = 2\pi$ and the curve

$$= \Delta = \frac{1}{2}a\int_0^{2\pi} r^2 d\theta$$

$$= \frac{1}{2}a^2\int_0^{2\pi} \tanh^2 \frac{1}{2}\theta d\theta$$

$$= \frac{1}{2}a^2 \int_0^{2\pi}\left(1 - \sec h^2 \frac{1}{2}\theta\right)d\theta$$

$$= \frac{1}{2}a^2\left[\theta - 2\tanh\frac{1}{2}\theta\right]_0^{2\pi}$$

$$= \frac{1}{2}a^2[2\pi - 2\tanh\pi]$$

$$= a^2\ [\pi - \tan h\ \pi]$$

$$= a\ [a\pi - a\tanh\pi]$$

$$= a\ [(2a\pi - a\tanh\pi) - a\pi]$$

$$= a\ (s - a\pi), \qquad \text{from (3).}$$

Example 60:

Prove the formula $s = \int \frac{r\,dr}{\sqrt{(r^2 - p^2)}}$.

Show that the arc of the curve $p^2\ (a^4 + r^4) = a^4 r^2$ *between the limits* $r = b$, $r = c$ *is equal in length of the arc of the hyperbola* $xy = a^2$ *between the limits* $x = b$, $x = c$.

Solution:

From differential calculus, we know that

$$\tan\phi = r\frac{d\theta}{dr} \text{ and } \frac{ds}{dr} = \sqrt{\left[1 + \left(r\frac{d\theta}{dr}\right)^2\right]}$$

$$\therefore \qquad \frac{ds}{dr} = \sqrt{(1 + \tan^2\phi)}$$

$$= \sqrt{(\sec^2\phi)} = \sec\phi$$

$$= \frac{1}{\cos\phi} = \frac{1}{\sqrt{(1 - \sin^2\phi)}}$$

$$= \frac{1}{\sqrt{\{1 - (p^2/r^2)\}}} \qquad [\because p = r\sin\phi]$$

$$= \frac{r}{\sqrt{(r^2 - p^2)}}.$$

Thus $$ds = \frac{r}{\sqrt{(r^2 - p^2)}}dr.$$

Integrating between the given limits, we get

$$s = \int \frac{r}{(r^2 - p^2)} dr. \qquad ...(1)$$

Now the given curve is $p^2 (a^4 + r^4) = a^4 r^2$

or $\qquad p^2 = a^4 r^2/(a^4 + r^2).$

We have $\qquad r^2 - p^2 = r^2 - \dfrac{a^4 r^2}{(a^4 + r^4)}$

$$= \frac{r^6}{(a^4 + r^4)}. \qquad ...(2)$$

Therefore from (1), the arc of the given curve between the limits $r = b$, $r = c$ is

$$= \int_b^c \frac{r\,dr}{\sqrt{(r^2 - p^2)}}$$

$$= \int_b^c \frac{r\,dr}{\sqrt{\{r^6 / (a^4 + r^4)\}}}, \qquad \text{from (2)}$$

$$= \int_b^c \frac{r\sqrt{(a^4 + r^4)}}{r^3} dr$$

$$= \int_b^c \frac{r\sqrt{(a^4 + r^4)}}{r^2} dr. \qquad ...(3)$$

Also for the hyperbola $xy = a^2$

i.e., $y = a^2/x$, $dy/dx = -a^2/x^2$.

$\therefore$ the arc length of the hyperbola $xy = a^2$ between the limits $x = b$, $x = c$

$$= \int_b^c \sqrt{\left\{1 + \left(\frac{dy}{dx}\right)^2\right\}} dx = \int_b^c \sqrt{\left\{1 + \frac{a^4}{x^4}\right\}} dx$$

$$= \int_b^c \frac{\sqrt{(x^4 + a^4)}}{x^2} dx = \int_b^c \frac{\sqrt{(r^4 + a^4)}}{r^2} dr,$$

changing the variable from x to r by a property of definite integrals

$$= \int_b^c \frac{\sqrt{(a^4 + r^4)}}{r^2} dr. \qquad ...(4)$$

From (3) and (4) we observe that the two lengths are equal.

Example 61:

Show that the intrinsic equation of the parabola $y^2 = 4ax$ is $s = a \cot\psi \operatorname{cosec}\psi + a \log(\cot\psi + \operatorname{cosec}\psi)$, ψ being the angle between the x-axis and the tangent at the point whose arcual distance from the vertex is s.

Solution:

The given parabola is $y^2 = 4ax$. ...(1)

Differentiating (1) w.r.t. x, we get $2y\,(dy/dx) = 4a$.

$\therefore \quad \tan\psi = dy/dx = 4a/2y = 2a/y$. ...(2)

If s denotes the arc length of the parabola measured from the vertex (0, 0) in the direction of y increasing, then

$$\frac{ds}{dy} = \sqrt{\left\{1+\left(\frac{dx}{dy}\right)^2\right\}}$$

$$= \sqrt{\left\{1+\frac{y^2}{4a^2}\right\}}, \qquad \left[\because \frac{dx}{dy} = \frac{y}{2a}\right]$$

$$= \sqrt{\left\{\frac{4a^2+y^2}{4a^2}\right\}} = \frac{1}{2a}\sqrt{(4a^2+y^2)}.$$

$$\therefore \qquad ds = \frac{1}{2a}\sqrt{(4a^2+y^2)}\,dy.$$

Integrating, $\displaystyle\int_0^s ds = \frac{1}{2a}\int_0^y \sqrt{\left(4a^2+y^2\right)}\,dy$

or $\displaystyle s = \frac{1}{2a}\left[\frac{1}{2}y\sqrt{\left(a^2+y^2\right)}+\frac{1}{2}\cdot 4a^2\log\left\{y+\sqrt{\left(4a^2+y^2\right)}\right\}\right]_0^y$

$$= (1/2a)\left[\frac{1}{2}y\sqrt{\left(4a^2+y^2\right)}+\frac{1}{2}\cdot 4a^2\log\left\{y+\sqrt{\left(4a^2+y^2\right)}\right\} - \frac{1}{2}\cdot 4a^2\log 2a\right]$$

$$= \frac{1}{4a}\left[y\sqrt{\left(4a^2+y^2\right)}+4a^2\log\frac{y+\sqrt{(4a^2+y^2)}}{2a}\right] \qquad ...(3)$$

Now to obtain the intrinsic equation of the given parabola we eliminate y between (2) and (3). From (2), we have $y = 2a\cot\psi$.

Putting this value of y in (3), we get

$$s = \frac{1}{4a}\left[2a\cot\psi\sqrt{(4a^2+4a^2\cos^2\psi)}\right.$$

$$\left. +4a^2\log\frac{2a\cot\psi+\sqrt{(4a^2+4a^2\cot^2\psi)}}{2a}\right]$$

$$= \frac{1}{4a}\left[(2a\cot\psi)\cdot 2a\sqrt{(1+\cot^2\psi)}\ +4a^2\log\left\{\cot\psi+\sqrt{(1+\cot^2\psi)}\right\}\right]$$

$= a\cot\psi\ \text{cosec}\ \psi + a\log(\cot\psi + \text{cosec}\ \psi)$, which is the required intrinsic equation.

Example 62:

prove that Prove that the intrinsic equation of the parabola $x^2 = 4ay$ is $s = a\tan\psi\sec\psi + a\log(\tan\psi + \sec\psi)$.

Solution:

We get $\tan\psi = dy/dx = x/2a$. ...(1)

$$\text{Also}s = \int_0^x\sqrt{\left\{1+\left(\frac{dy}{dx}\right)^2\right\}}dx = \frac{1}{2a}\int_0^x\sqrt{(4a^2+x^2)}dx$$

$$= \frac{1}{4a}\left[x\sqrt{(4a^2+x^2)}+4a^2\log\frac{x+\sqrt{(x^2+4a^2)}}{2a}\right], \quad ...(2)$$

Eliminating x from (1) and (2), we get $s = a[\tan\psi\sec\psi + \log(\tan\psi + \sec\psi)]$,

which is the required intrinsic equation.

Example 63:

Find the intrinsic equation of the parabola $y^2 = 4ax$. Hence deduce the length of the arc measured from the vertex to an extremity of the latus rectum.

Solution:

We have already obtained the intrinsic equation of the parabola $y^2 = 4ax$

$$s = a[\text{cosec}\ \psi\cot\psi + \log(\text{cosec}\ \psi + \cot\psi), \quad ...(1)$$

where ψ is the angle between the x-axis and tangent at the point whose arcual distance from the vertex is s.

Now in the intrinsic equation (1) of the parabola the arc length s has been measured from the vertex. We want to find the length of the arc from the vertex to an extremity of the latus rectum Let this length be s_1.

At an extremity of the latus rectum, $y = 2a$. Also $\tan\psi = y/2a$. So at an extremity of the latus rectum, $\tan\psi = 2a/2a = 1$ *i.e.*, $\psi = \pi/4$.

So putting $\psi = \pi/4$ in (1), we get

$$s_1 = a\left[\operatorname{cosec}\frac{1}{4}\pi\cot\frac{1}{4}\pi + \log\left(\operatorname{cosec}\frac{1}{4}\pi + \cot\frac{1}{4}\pi\right)\right]$$
$$= a\,[\sqrt{2} + \log(1 + \sqrt{2})].$$

Example 64:

Show that the intrinsic equation of the semi-cubical parabola $3ay^2 = 2x^3$ is $9s = 4a(\sec^3\psi - 1)$.

Solution:

The given semicubical parabola is $3ay^2 = 2x^3$. ...(1)

Differentiating (1) w.r.t. x, we get $6ay\,(dy/dx) = 6x^2$

or
$$\frac{dy}{dx} = \frac{x^2}{ay} = \frac{x^2}{a\sqrt{(2x^3/3a)}} = \sqrt{\left(\frac{3x}{2a}\right)}.$$

$\therefore$
$$\tan\psi = \frac{dy}{dx} = \sqrt{\left(\frac{dx}{2a}\right)}. \quad ...(2)$$

If s denotes the arc length of the given curve measured from the point (0, 0) to any point P (x, y) in the direction of x increasing, then

$$s = \int_0^x \sqrt{\left\{1 + \left(\frac{dy}{dx}\right)^2\right\}}dx = \int_0^x \sqrt{\left(1 + \frac{3x}{2a}\right)}dx$$

$$= \int_0^x \left(1 + \frac{3x}{2a}\right)^{1/2} dx = \left[\frac{\{1 + (3x/2a)\}^{3/2}}{(3/2)\cdot(3/2a)}\right]_0^x$$

$$= \frac{4a}{9}\left[\left(1 + \frac{3x}{2a}\right)^{3/2} - 1\right] \quad ...(3)$$

Eliminating x between (2) and (3), we get

$$s = \frac{4a}{9}\left[\left(1 + \tan^2\psi\right)^{3/2} - 1\right] = \frac{4a}{9}\left(\sec^3\psi - 1\right),$$

which is the required intrinsic equation of the curve.

Example 65:

Show that the intrinsic equation of $ay^2 = x^3$ taking its cusp as the fixed point $27s = 8a(\sec^3\psi - 1)$.

Solution:

The given curve is $ay^2 = x^3$. ...(1)

We get

$$\tan\psi = \frac{dy}{dx} = \frac{3}{2}\frac{1}{\sqrt{a}}x^{1/2} \qquad ...(2)$$

and
$$s = \int_0^x \sqrt{\left(1+\frac{9x}{4a}\right)}dx$$

$$= \frac{1}{27\sqrt{a}}\left[(4a+9x)^{3/2} - 8a^{3/2}\right]. \qquad ...(3)$$

Eliminating x between (2) and (3), we get $27s = 8a(\sec^3\psi - 1)$.

Example 66:

Find the intrinsic equation of the catenary $y = c\cos h\,(x/c)$. Hence show that $c\rho = c^2 + s^2$, where r is the radius of curvature.

Solution:

The given curve is $y = c\cos h\,(x/c)$. ...(1)

Differentiating (1) w.r.t. x, we get $dy/dx = c\sin h\,(x/c).(1/c) = \sin h\,(x/c)$.

$$\therefore \qquad \tan\psi = dy/dx = \sin h\,(x/c). \qquad ...(2)$$

If s denotes the arc length of the catenary measured from the vertex (0, c) to any point (x, y) in the direction of x increasing, then

$$s = \int_0^x \sqrt{\left\{1+\left(\frac{dy}{dx}\right)^2\right\}}dx = \int_0^x \sqrt{\left\{1+\sinh^2\frac{x}{c}\right\}}dx$$

$$= \int_0^x \cosh\frac{x}{c}dx = c\sin h\,\frac{x}{c}. \qquad ...(3)$$

Eliminating x between (2) and (3), we get $s = c\tan\psi$, which is the equation of the catenary.

Also
$$\rho = \frac{ds}{d\psi} = c\sec^2\psi = c(1+\tan^2\psi) = c\left(1+\frac{s^2}{c^2}\right)$$

or
$$c\rho = c^2 + s^2.$$

Example 67:

Show that the intrinsic equation of the cycloid $x = a(t + \sin t)$, $y = a(1 - \cos t)$ is

$$s = 4a\sin\psi.$$

Hence or otherwise find the length of the complete cycloid.

Solution:

The given equation of the cycloid are

$$x = a(t + \sin t),\ y = a(1 - \cos t). \qquad ...(1)$$

We have $dx/dt = a(1 + \cos t)$, and $dy/dt = a \sin t$.

$$\therefore \qquad \frac{dy}{dx} = \frac{dy/dt}{dx/dt} = \frac{a \sin t}{a(1+\cos t)}$$

$$= \frac{2\sin\frac{1}{2}t\cos\frac{1}{2}t}{2\cos^2\frac{1}{2}t} = \tan\frac{1}{2}t.$$

Hence $\tan\psi = dy/dx = \tan\frac{1}{2}t$ or $\psi = \frac{1}{2}t$. ...(2)

If s denotes the arc length of the cycloid measured from the vertex (*i.e.*, the point $t = 0$) to any point P (*i.e.*, the point 't') in the direction of t increasing, then

$$s = \int_0^t \sqrt{\left\{\left(\frac{dx}{dt}\right)^2 + \left(\frac{dy}{dt}\right)^2\right\}}dt$$

$$= \int_0^t \sqrt{\left\{a^2(1+\cos t)^2 + a^2\sin^2 t\right\}}dt$$

$$= \int_0^t \sqrt{\left\{2a^2(1+\cos t)\right\}}dt$$

$$= 2a\int_0^t \cos\frac{1}{2}t\,dt = 2a\left[2\sin\frac{1}{2}t\right]_0^t = 4a\sin\frac{1}{2}t \qquad ...(3)$$

Eliminating t from (2) and (3), we get $s = 4a\sin\psi$, ...(4)

which is the required intrinsic equation of the cycloid.

Second Part

In the intrinsic equation (4) of the cycloid the arc length s has been measured from the vertex *i.e.*, the point $\psi = 0$ At a cusp, we have $t = \pi$ and $\psi = \pi/2$. If s_1 denotes the length of the arc extending from the vertex to a cusp, then from (4), we have $s_1 = 4a \sin 1/2\pi = 4a$.

$\therefore$ the whole length of an arch of the cycloid $= 2 \times 4a = 8a$.

Example 68:

Prove that the intrinsic equation of the curve $x = a(1 + \sin t)$, $y = a(1 + \cos t)$ is $s + a\psi = 0$.

Solution:

The given curve is $x = a(1 + \sin t)$, $y = a(1 + \cos t)$. ...(1)

$\therefore$ $dx/dt = a\cos t$ and $dy/dt = -a\sin t$.

$$\therefore \qquad \frac{dy}{dx} = \frac{dy/dt}{dx/dt}$$

$$= \frac{-a\sin t}{a\cos t} = -\tan t.$$

Hence $\qquad \tan\psi = dy/dx$

$$= -\tan t = \tan(-1)$$

so that $\qquad \psi = -t.$ ...(2)

Measuring the arc length s from the point t = 0, we have

$$s = \int_0^t \sqrt{\left\{\left(\frac{dx}{dt}\right)^2 + \left(\frac{dy}{dx}\right)^2\right\}}dt$$

$$= \int_0^t \sqrt{\left(a^2\cos^2 t + a^2\sin^2 t\right)}dt = a\int_0^t dt \quad ...(3)$$

Eliminating t from (2) and (3), the intrinsic equation is

$$s = a(-\psi) \text{ or } s + a\psi = 0.$$

Example 69:

In the four-cusped astroid $x^{2\ 3} + y^{2\ 3} = a^{2\ 3}$, show that

(i) $s = \dfrac{3}{4} a\cos^2\psi$, *s being measured from the vertex;*

(ii) $s = \dfrac{3}{2} a\sin^2\psi$, *s being measured from the cusp on x-axis;*

(iii) whole length of the curve is 6a.

Solution:

The parametric equation of the given curve are

$$x = a\cos^3 t,\ y = a\sin^3 t. \quad ...(1)$$

We have $\qquad dx/dt = -3a\cos^2 t\sin t$, a

and $\qquad dy/dt = 3a\sin^2 t\cos t.$

$$\therefore \qquad \frac{dy}{dx} = \frac{dy/dt}{dx/dt} = \frac{3a\sin^2 t\cos t}{-3a\cos^2 t\sin t} = -\tan t.$$

Sol we have $\tan\psi = dy/dx = -\tan t = \tan(-t).$

$\therefore \qquad \psi = -t.$ ...(2)

Now $\qquad \left(\dfrac{ds}{dt}\right)^2 = \left(\dfrac{dx}{dt}\right)^2 + \left(\dfrac{dy}{dt}\right)^2$

$$= (9a^2\cos^4 t\sin^2 t) + (9a^2\sin^4 t\cos^2 t)$$

$$= 9a^2 \sin^2 t \cos^2 t (\cos^2 t + \sin^2 t)$$

$$= 9a^2 \sin^2 t \cos^2 t. \qquad ...(3)$$

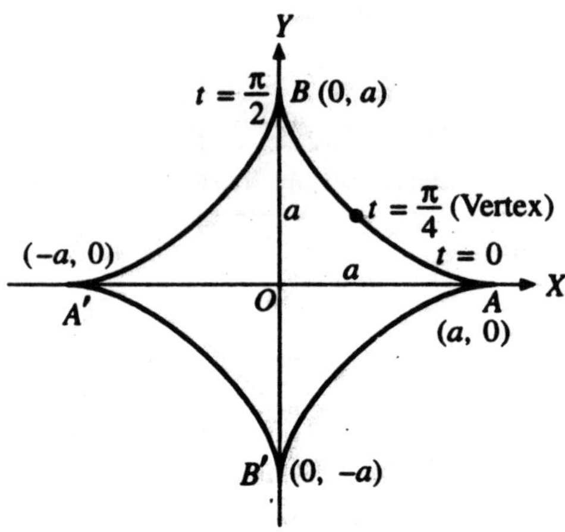

Fig. 2.14

(i) If s denotes the arc length of the given curve measured from the vertex (*i.e.*, the middle point of the arc in the Ist quadrant) to any point P lying towards the cusp on x-axis, then s increases as t decreases. Therefore ds/dt will be negative, so from (3), we have

$$ds/dt = -\ 3a \sin t \cos t$$

or
$$ds = -\ 3a \sin t \cos t\ dt. \qquad ...(4)$$

Now at the vertex of the given curve, we have $t = \pi/4$.

∴ from (4), the arcual distance s measured from the vertex is given by

$$s = -\ 3a \int_{\pi/4}^{t} \sin t \cos dt = -\ \frac{3a}{2} \int_{\pi/4}^{t} \sin 2t\, dt$$

$$= -\ \frac{3a}{2} \left[-\frac{\cos 2t}{2} \right]_{\pi/4}^{t}$$

$$= \frac{3}{4} a \cos 2t. \qquad ...(5)$$

Eliminating t between (2) and (5), the required intensity equation of the curve is

$$s = \frac{3}{4}\ a \cos \{2 (-\ \psi)\}$$

$$= \frac{3}{4}\ a \cos 2\ \psi. \qquad [\because \cos(-\theta) = \cos\theta]$$

(ii) If s denotes the arc length of the given curve measured from the cusp on x-axis to any point P lying towards the second cusp on y-axis, then s increases as t increases. Therefore ds/dt will be positive. Hence from (3), we have

$$ds/dt = 3a \sin t \cos t \text{ or } ds = 3a \sin t \cos t \, dt.$$

Also at the cusp on x-axis, we have t = 0.

$$\therefore \qquad s = \int_0^t 3a \sin t \cos t \, dt = 3a \left[\frac{\sin^2 t}{2} \right]_0^t$$

$$= \frac{3}{2} a \sin^2 t.$$

Eliminating t between (2) and (6), the required intrinsic equation of the curve is

$$s = \frac{3}{2} a \sin^2 (-\psi)$$

or

$$s = \frac{3}{2} a \sin^2 \psi.$$

(iii) The whole length of the curve is already obtained.

3

Volumes and Surfaces of Solids of Revolution

3.1 Definition

Solid of Revolution : If a plane area is revolved about a fixed line in its own plane, then the body so generated by the revolution of the plane area is called a solid of revolution.

Surface of Revolution : If a plane curve is revolved about a fixed line lying in its own plane, then the surface generated by the perimeter of the curve is called a surface of revolution.

Axis of Revolution : The fixed straight line, say AB, about which the area revolves is called the axis of revolution or axis of rotation.

3.2 Volumes of Solids of Revolution

(a) The Axis of Rotation being x-axis : If a plane area bounded by the curve $y = f(x)$, the ordinates $x = a$, $x = b$ and the x-axis revolves about the x-axis then the volume of the solid thus generated is

$$\int_a^b \pi y^2 dx = \int_a^b \pi [f(x)]^2 dx,$$

where $y = f(x)$ is a finite, continuous and single valued function of x in the interval $a \le x \le b$.

Or

The volume of the solid generated by the revolution of the area bounded by the curve $y = f(x)$, x-axis and the ordinates $x = a$, $x = b$ about the x-axis is $\int_a^b \pi y^2 dx$.

Proof:

Let AB be the arc of the curve $y = f(x)$ included between the ordinates $x = a$ and $x = b$. It is being assumed that the curve does not cut the x-axis and $f(x)$ is a continuous function of x in the travel (a, b).

Let P (x, y) and Q (x + δx, y + δy) be any two neighboring point on the curve y = f(x). Draw the ordinates PM and QN. Also draw PP' and QQ perpendiculars to these ordinates.

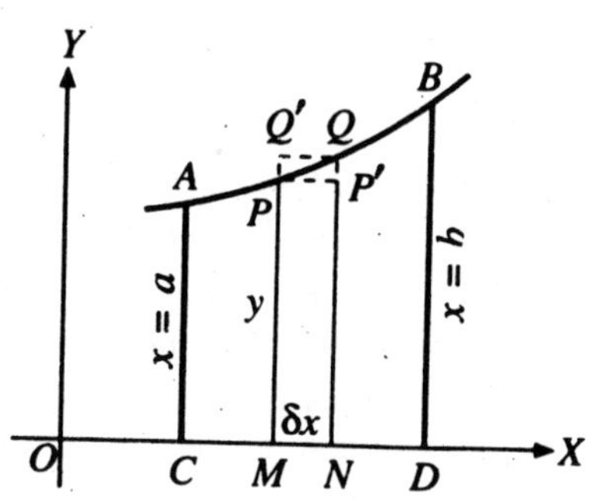

Fig. 3.1

Let V denote the volume of the solid generated by the revolution of the area ACMP about the x-axis and let the volume of revolution obtained by revolving the area ACNQ about x-axis be V + δV, so that volume of the solid generated by the revolution of the strip PMNQ about the x-axis is δV.

Now PM = y, QN = y + δy and MN = (x + δx) – x = δx. Then the volume of the solid generated by revolving the area PMNP' = $\pi y^2 \delta x$ and the volume of the solid generated by revolving the area Q'MNQ = $\pi (y + \delta y)^2 \delta x$.

Also the volume of the solid generated by the revolution of the area PMNQP (*i.e.*, the volume δV) lies between the volumes of the right circular cylinders generated by the revolution of the areas PMNP'P and MNQQ' *i.e.*, δV lies between

$$\pi y^2 \delta x \text{ and } \pi (y + \delta y)^2 \delta x$$

or $\quad (\delta V/\delta x)$ lies between πy^2 and $\pi (y + \delta y)^2$

i.e., $\quad \pi y^2 < (\delta V/\delta x) < \pi (y + \delta y)^2.$

In the limiting position Q → P, δx → 0 (and therefore δy → 0), we have

$$dV/dx = \pi y^2 \text{ or } dV = \pi y^2 dx.$$

Hence $\quad \int_a^b \pi y^2 dx = \int_a^b dV = [V]_{x=a}^{x=b}$

= (value of V for x = b) – (value of V for x = a)

= volume generated by the area ACDB – 0

= volume of the solid generated by the revolution of the revolution of the given area ACDB about the axis of x.

∴ the required volume = $\pi \int_a^b y^2 \, dx$.

(b) The Axis of Rotation being y-axis : Similarly, it can be shown that the *volume of the solid generated by the revolution about y-axis of the area*

between the curve $x = f(y)$, the y-axis and the two abscissae $y = a$ and $y = b$ is given by

$$\int_a^b \pi\, x^2 dy.$$

(c) Volume of Solid of Revolution When the Equations of the Generating Curve are Given in Parametric Form

(i) If the curve is given by the parametric equation, say $x = \phi(t)$, $y = \psi(t)$, then the volume of the solid generated by the revolution about x-axis of the area bounded by the curve, the axis of x and ordinate at the points where $t = a$ and $t = b$ is

$$= \int_a^b \pi\, y^2 \frac{dx}{dt} dt = \pi \int_a^b \{\psi(t)\}^2 \phi'(t) dt.$$

(ii) The volume of the solid generated by the revolution about y-axis of the area between the curve $x = \phi(t)$, $\psi = y(t)$, the y-axis and the abscissan at the point where $t = a$, $t = b$ is

$$= \int_a^b \pi\, x^2 \frac{dy}{dt} dt = \pi \int_a^b \{\phi(t)\}^2 \psi'(t) dt.$$

(d) Volume of Solid of Revolution When the Equation of the Generating Curve is Given in Polar Co-ordinates : If the equation of the generating curve is given in polar co-ordinates, say $r = f(\theta)$, and the curve revolves about the axis of x, the volume generated

$$= \pi \int_{x=a}^{b} y^2 dx = \pi \int_{\theta=\alpha}^{\beta} y^2 \frac{dx}{d\theta} d\theta,$$

where α and β are the values of θ at the points where $x = a$ and $x = b$ respectively.

Now $x = r\cos\theta$ and $y = r\sin\theta$. Therefore the volume

$$= \pi \int_{\theta=\alpha}^{\beta} r^2 \sin^2\theta \frac{d}{d\theta}(r\cos\theta) d\theta,$$

in which the value of r in terms of θ must be substituted from the equation of the curve.

A similar procedure can be adopted in case the curve revolves about the axis of y.

Alternative Method in the Case of Polar Curve

The volume of the solid generated by the revolution of the area bounded by the curve $r = f(\theta)$ and radii vectors $\theta = \theta_1$, $\theta = \theta_2$.

(i) about the initial line $\theta = 0$ (i.e., the x-axis) is

$$\int_{\theta_1}^{\theta_2} \frac{2}{3}\pi r^2 \sin\theta \, d\theta,$$

(ii) *about the line* $\theta = \theta/2$ *(i.e., the y-axis) is*

$$\int_{\theta_1}^{\theta_2} \frac{2}{3}\pi r^3 \cos\theta \, d\theta,$$

(iii) *about any line* $(\theta = \gamma)$ *is*

$$\int_{\theta_1}^{\theta_2} \frac{2}{3}\pi r^3 \sin(\theta - \gamma) d\theta,$$

where in each of the above three formulae the value of r in terms of θ must be substituted from the equation of the given curve.

3.3 Surfaces of Solids of Revolution

(a) *Revolution About the Axis of x : To prove that the curved surface of the solid generated by the revolution, about x-axis, of the area bounded by the curve* $y = f(x)$*, the ordinates* $x = a$*,* $x = b$ *and the x-axis is*

$$\int_{x=a}^{x=b} 2\pi y \, ds$$

where s is the length of the arc measured from $x = a$ *to any point (x, y).*

Or

Show that the area of the surface of the solid obtained by revolving about x-axis the arc of the curve intercepted between the points whose abscissa are a and b is

$$\int_a^b 2\pi y \frac{ds}{dx} dx$$

Proof:

Let AB be the arc of the curve $y = f(x)$ included between the ordinates $x = a$ and $x = b$. It is being assumed that the curve does not cut x-axis and $f(x)$ is a continuous function of x in the interval (a, b).

Let P (x, y) and Q $(x + \delta y, y + \delta y)$ be any two neighbouring points on the curve $y = f(x)$.

Let the length of the arc AP be s and arc AQ $= s + \delta s$ so that arc PQ $= \delta s$.

Draw the ordinates PM and QN. Let S denote the curved surface of the solid generated by the revolution of the area CMPA about the x-axis. Then the curved surface of the solid generated by the revolution of the area MNQP $= \delta S$.

We shall take it as an axiom that the curved surface of the solid generated by the revolution of the area MNQP about the x-axis lies between the curved surfaces of the right circular cylinders whose radii are PM and NQ and which are of the same thickness (height) δs. There is no loss in assuming so because ultimately Q is to tend to P.

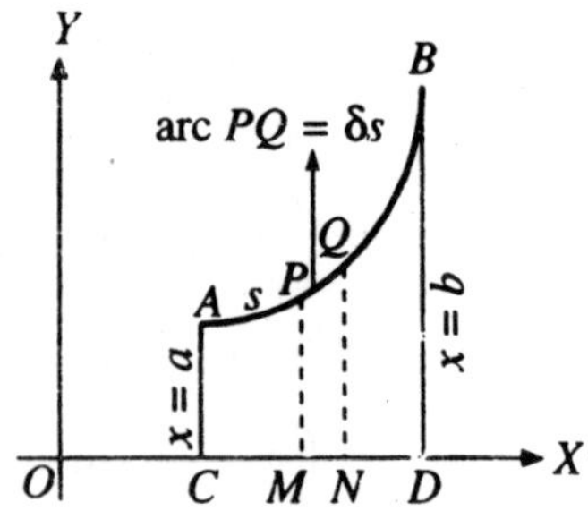

Fig. 3.2

Thus, δS lies between 2π y δs and 2π (y + δy) δs

i.e., $$2\pi y\, \delta s < \delta S < 2\pi (y + \delta y)\, \delta s$$

or $$2\pi y < (\delta/\delta s) < 2\pi (y + \delta y).$$

Now as Q approaches P *i.e.*, δs → 0, dy will also tend to zero. Hence by taking limits as δs → 0, we have

$$\frac{dS}{ds} = 2\pi y \text{ or } dS = 2\pi y\, ds.$$

$$\therefore \quad \int_{x=a}^{x=b} 2\pi y\, ds = \int_{x=a}^{x=b} dS = [S]_{x=a}^{x=b}$$

= (the value of S when x = b) – (the value of S when x = a)

= surface of the solid generated by the revolution of the area ACDB – 0.

$$\therefore \quad \text{the required cureved surface} = \int_{x=a}^{x=b} 2\pi y\, ds$$

$$= \int_{x=a}^{x=b} 2\pi y \frac{ds}{dx} dx, \text{ where } \frac{ds}{dx} = \sqrt{\left\{1 + \left(\frac{dy}{dx}\right)^2\right\}}$$

(b) *Axis of Revolution as y-axis* : Similarly, the curved surface of the solid generated by the revolution about the y-axis, of the area bounded by the curve x = f (y), the lines y = a, y = b and the y-axis is

$$2\pi \int_{x=a}^{x=b} x\, ds$$

or $$S = 2\pi \int_{y=a}^{b} x \frac{ds}{dy} dy, \text{ where } \frac{ds}{dy} = \sqrt{\left\{1 + \left(\frac{dx}{dy}\right)^2\right\}}.$$

(c) *Surface Formula for Parametric Equations* : Suppose the equation of the curve is given in parametric form x = f (x), y = φ(t), t being the variable parameter. Then the curved surface of the solid formed by the revolution about the x-axis

$$= \int 2\pi y \frac{ds}{dt} dt, \text{ between the suitable limits}$$

where $$\frac{ds}{dt} = \sqrt{\left\{\left(\frac{dx}{dt}\right)^2 + \left(\frac{dy}{dt}\right)^2\right\}}.$$

(d) *Surface Formula for Polar Equations :* Suppose the equation of the curve is given in the polar form $r = f(\theta)$. Then the curve surface generated by the revolution about the initial line, of the arc intercepted between the radii vectors $\theta = a$ and $\theta = \beta$ is

$$\int_{\theta=\alpha}^{\theta=\beta} 2\pi (r \sin\theta) \frac{ds}{d\theta} d\theta, \text{ where } \frac{ds}{d\theta} = \sqrt{\left\{r^2 + \left(\frac{dr}{d\theta}\right)^2\right\}}. \quad [y = r \sin\theta]$$

Note: In some cases we may use the formula

$$S = \int 2\pi y \frac{ds}{dr} dr, \text{ where } \frac{ds}{dr} = \sqrt{\left\{1 + \left(r \frac{d\theta}{dr}\right)^2\right\}}.$$

(e) *Revolution About Any Axis.* If the given arc AB is revolved about a line CD other than the coordinate axes, then the curved surface thus generated is

$= 2\pi \int (PM) ds$, (between the proper limits of integration), where PM is the perpendicular drawn from any point P on the arc AB to the axis of revolution CD and ds is the length of an element of the arc AB at the point P.

Remark:

If an arc length revolves about x-axis, the basic formula for the surface of revolution in all cases is $\int 2\pi y \, ds$, between the suitable limits. If we want to integrate w.r.t. x, we shall change ds as (ds/dx) dx and adjust the limits accordingly.

A similar transformation can be made if we want to integrate w.r.t. y or with respect to θ or w.r.t. Some parameter, say t.

3.4 Theorems of Pappus or Guldin

Theorem 1:

Surface of a Solid of Revolution : *If an arc of a plane curve revolves about a strength line in its place, which does not intersect it, the surface of*

the solid thus obtained is equal to the arc multiplied by the length of the path described by the centroid of the arc.

Proof:

Let l be the length of the arc AB and let it revolve about OX.

Let the absassae of the extremities A and B of the arc be a and b.

Then the surface generated by the revolution of the arc AB about x-axis is

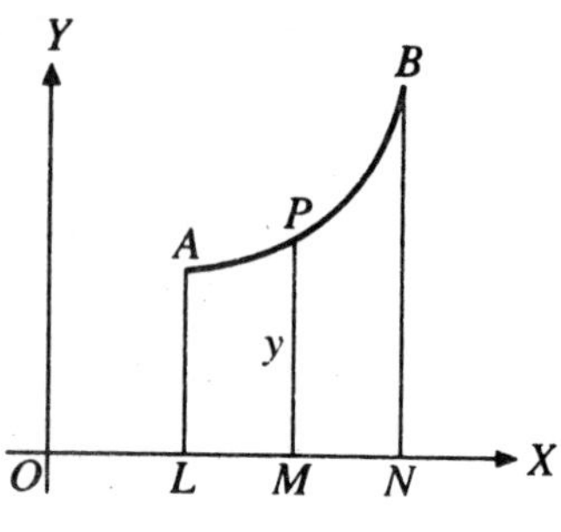

Fig. 3.3

$$= \int_{x=a}^{x=b} 2\pi\, y ds \qquad ...(1)$$

Also we know that (see the chapter on centre of gravity) the ordinate y, of the centroid of the arc from x = a to x = b, of length l, is given by

$$\bar{y} = \frac{\int_{x=a}^{b} y\, ds}{l}$$

From (1) and (2), we get the required surface

$= 2\pi \bar{y}_1 = l \times 2\pi \bar{y}$

= length of the arc × length of the path described by the centroid of the arc.

Notes:

1. The closed curve or arc in the above theorems must not cross the axis of revolution but may be terminated by it.
2. When the volume or surface generated is known, the theorems may be applied to find the position of the centroid of the generating area of arc.

Ex. *State and prove the theorems of Pappus and Guldin.*

Theorem 2:

Volume of a Solid of Revolution: *If a closed plane curve revoles about a straight line in its plane which does not intersect it, the volume of the ring thus obtained is equal to the area of the region enclosed by the curve multiplied by the length of the path described by the centroid of the region.*

Proof:

Let AP_1BP_2A be the closed plane curve and let it rotate about the axis of x.

Let AL (x = a) and BN (x = b) be the tangents to the curve parallel to the y-axis (a < b). Also let any ordinate meet the curve at P_1, P_2 and let $MP_1 = y_1$, $MP_2 = y_2$ so that y_1, y_2 are functions of x.

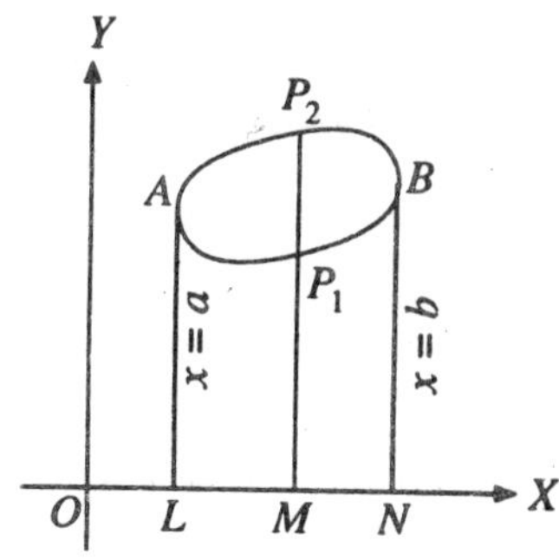

Fig. 3.4

Now volume of the ring generated by the revolution of the closed curve AP_1BP_2A about the axis of x = volume generated by the area $ALNBP_2A$ volume by the area $ALNBP_1A$

$$= \pi\int_a^b y_2^2 dx - \pi\int_a^b y_1^2 dx$$

$$= \pi\int_a^b (y_2^2 - y_1^2) dx. \qquad ...(1)$$

Also if $\bar{y}$ be the ordinate of the centroid of the area of the closed curve, then

$$\bar{y} = \frac{\int_a^b \frac{1}{2}(y_1 + y_2)(y_2 - y_1)dx}{A}$$

$$= \frac{\frac{1}{2}\int_a^b (y_2^2 - y_1^2)dx}{A}, \qquad ...(2)$$

where A is the area of the closed curve.

Hence form (1) and (2), the required volume

$= 2\pi A\bar{y} = A \times 2\pi\bar{y}$

= area of the closed curve ×circumference of the circle of radius $\bar{y}$

= area of the curve × length of the arc described by the centroid of the region bounded by the closed curve.

MISCELLANEOUS EXAMPLES

Example 1:

The arc of the cardioid $r = a(1 + \cos\theta)$, *specified by* $-\frac{1}{2}\pi < \theta < \frac{1}{2}\pi$ *is rotated about the line* $\theta = 0$, *prove that the area of the surface generated is* $\frac{4}{5}(8 - \sqrt{2})\pi a^2$.

Solution:

Obviously the surface generated by revolving the arc lying between $-\frac{1}{2}\pi \leq \theta \leq \frac{1}{2}\pi$ about the initial line is the same as that generated by revolving the arc lying between between $\theta = 0$ and $\theta = \frac{1}{2}\pi$.

$$\therefore \quad \text{the required surface} = \int_0^{\pi/2} 2\pi\, y \frac{ds}{d\theta} d\theta$$

$$= 16\pi a^2 \int_0^{\pi/2} \cos^4 \frac{\theta}{2} \sin \frac{\theta}{2} d\theta,$$

$$= 16\pi a^2 \left[\frac{-\cos^5 \frac{1}{2}\theta}{5 \cdot \frac{1}{2}} \right]_0^{\pi/2}$$

$$= -\frac{32}{5}\pi a^2 [(1/\sqrt{2})^5 - 1]$$

$$= \frac{32\pi a^2}{5}\left(1 - \frac{1}{4\sqrt{2}}\right)$$

$$= \frac{4\pi a^2}{5}\left(8 - \frac{8}{4\sqrt{2}}\right)$$

$$= \frac{4\pi a^2}{5}(8 - \sqrt{2}).$$

Example 2:

The arc of the cardioid $r = a\,(1 + \cos\theta)$ *included between* $-\frac{1}{2}\pi \leq \theta \leq \frac{1}{2}\pi$ *is rotated about the line* $\theta = \frac{1}{2}\pi$. *Find the area of the surface generated.*

Solution:

The given cardioid is $r = a\,(1 + \cos\theta)$, ...(1)

which is symmetrical about the initial line.

$ds/d\theta = 2a \cos(\theta/2)$.

The curve is rotated about the line $\theta = \pi/2$ *i.e.*, the y-axis and for the upper half of the cardioid lying in the first quadrant, θ varies from θ to $\frac{1}{2}\pi$.

$$\therefore \quad \text{the required surface} = \int_0^{\pi/2} 2\pi\, x \frac{ds}{d\theta} d\theta,$$

$$= 4\pi \int_0^{\pi/2} r\cos\theta . 2a\cos\frac{\theta}{2}\, d\theta, \qquad [\because x = r\cos\theta]$$

$$= 8\pi a \int_0^{\pi/2} a(1+\cos\theta).\cos\theta.\cos\frac{\theta}{2}\, d\theta, \quad \text{from (1)}$$

$$= 8\pi a^2 \int_0^{\pi/2} 2\cos^2\frac{\theta}{2}\cdot\left(1-2\sin^2\frac{\theta}{2}\right)\cos\frac{\theta}{2}\, d\theta$$

$$= 16\pi a^2 \int_0^{\pi/2} \left(1-\sin^2\frac{\theta}{2}\right)\left(1-2\sin^2\frac{\theta}{2}\right)\cos\frac{\theta}{2}\, d\theta$$

$$= 16\pi a^2 \int_0^{\pi/2} \left(1-3\sin^2\frac{\theta}{2}+2\sin^4\frac{\theta}{2}\right)\cos\frac{\theta}{2}\, d\theta.$$

Put $\sin(\theta/2) = t$

so that $\frac{1}{2}\cos(\theta/2)\, d\theta = dt$.

Also when $\theta = 0$, $t = 0$

and when $\theta = \pi/2$, $t = 1/\sqrt{2}$.

$\therefore$ the required surface

$$= 16\pi a^2 \int_0^{1/\sqrt{2}} (1-3t^2+2t^4).2\, dt = 32\pi a^2\left[t - t^3 + 2\cdot\frac{t^5}{5}\right]_0^{1/\sqrt{2}}$$

$$= 32\pi a^2\left[\frac{1}{\sqrt{2}} - \frac{1}{2\sqrt{2}} + \frac{1}{10\sqrt{2}}\right] = \frac{48\sqrt{2}}{5}\pi a^2.$$

Example 3:

Prove that the volume of the solid generated by the revolution of an ellipse round its minor axis is a mean proportional between those generated by the revolution of the ellipse and of the auxiliary circle about the major axis.

Solution:

Let the ellipse be $x^2/a^2 + y^2/b^2 = 1$. ...(1)

Also the equation of its auxiliary circle is $x^2 + y^2 = a^2$. ...(2)

Now the volume of the solid generated by the revolution of (1) about the major axis *i.e.*, the axis of x is $4/3\pi ab^2 = V_1$, say.

Also the volume of the sphere formed by the revolution of (2) about x-axis is $\frac{4}{3}\pi a^3 = V_2$, say.

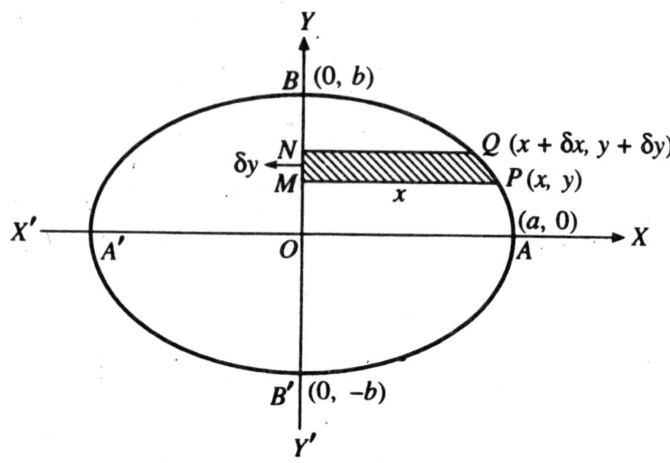

Fig. 3.5

Now we have to find the volume of the solid formed by the revolution of the ellipse about the minor axis *i.e.*, the axis of y. Consider an elementary strip PMNQ perpendicular to y-axis. Then the volume of the elementary disc formed by revolving the strip PMNQ about y-axis = $\pi x^2 \delta y$.

The ellipse is symmetrical about x-axis and on the arc AB of the ellipse y varies from 0 to b.

∴ the volume of the solid generated by the revolution of the ellipse about y-axis $= 2\int_0^b \pi x^2 dy$

$$= 2\pi \frac{a^2}{b^2}\int_0^b (b^2 - y^2)dy,$$

$$\because \text{ from (1), } x^2 = \frac{a^2}{b^2}(b^2 - y^2)$$

$$= \frac{2\pi a^2}{b^2}\left[b^2 y - \frac{y^3}{3}\right]_0^b$$

$$= \frac{2\pi a^2}{b^2}\left[b^3 - \frac{b^3}{3}\right] = \frac{4\pi a^2 b}{3} V_3, \text{ say.}$$

Now mean proportional between V_1 and V_2

$$= \sqrt{(V_1\ V_2)} = \sqrt{\left(\frac{4}{3}\pi ab^2 \cdot \frac{4}{3}\pi a^3\right)} = \frac{4}{3}\pi a^2 b = V_2$$

=volume generated when ellipse is revolved about minor axis.

Example 4:

Find the volume of the solid generated by the revolution of an arc of the catenary y = ccosh (x/c) about the x-axis.

Solution:

The given equation of catenary is y = ccosh (x/c). Let AL be an arc of this catenary where L is the point (x, y).

Take an elementary strip PMNQ perpendicular to the axis of x, so that PM = y and MN = dx.

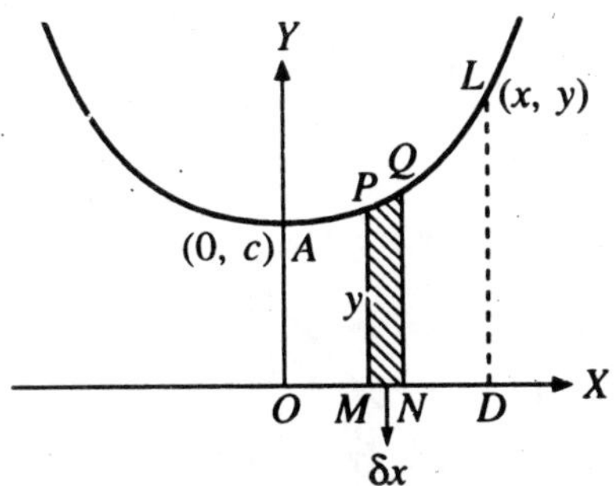

Fig. 3.6

Now volume of the elementary disc formed by revolving the strip PMNQ about the axis of x = π. PM^2. MN = πy^2dx.

$$\therefore \quad \text{the required volume} = \int_0^x \pi y^2 dx$$

$$= \int_0^x c^2 \cosh \frac{x}{c} dx, \qquad [\because\ y = \cosh (x/c)]$$

$$= \frac{\pi c^2}{2} \int_0^x \left(1 + \cosh \frac{2x}{c}\right) dx$$

$$= \frac{\pi c^2}{2} \left[x + \frac{c}{2} \sinh \frac{2x}{c}\right]_0^x$$

$$= \frac{\pi c^2}{2} \left[x + \frac{c}{2} \sinh \frac{2x}{c}\right].$$

The solid or revolution formed by revolving a catenary about its directrix is called a *catenoid.*

Example 5:

If the hyperbola $x^2/a^2 - y^2/b^2 = 1$ revolves about the x-axis, show that the volume included between the surface thus generated, the cone generated by the asymptotes and two planes perpendicular to the axis of x, at a distance h apart, is equal to that of a circular cylinder of height h and radius b.

Solution:

The given hyperbola is $x^2/a^2 - y^2/b^2 = 1$, ...(1)

and the equation of its asymptotes is $x^2/a^2 - y^2/b^2 = 0$

or $$y = \pm (b/a)\, x.$$

Let the two given planes be at distances c and (c + h) from the origin O. Then the volume of the portion of the cone generated by the asymptotes between x = c and x = c + h is

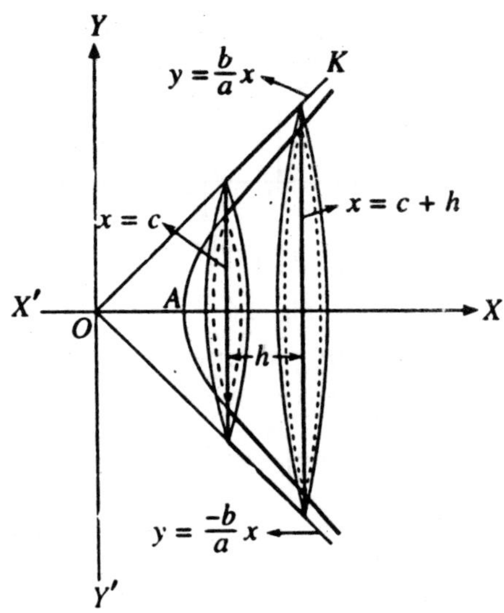

Fig. 3.7

$$= \int_c^{c+h} \pi y^2 dx,$$

where $$y = \frac{b}{a}x$$

$$= \pi \int_c^{c+h} \frac{b^2}{a^2} x^2 dx$$

$$= \frac{\pi b^2}{a^2}\left[\frac{x^3}{3}\right]_c^{c+h}$$

$$= \frac{\pi b^2}{2a^2}[(c+h)^3 - c^3]$$

$$= \frac{\pi b^2}{3a^2}[h^3 + 3ch^2 + 3c^2h] = V_1, \text{ say.}$$

Now the volume of the portion of the solid generated by the hyperbola between the two given places (*i.e.*, between x = c and x = c + h) is

$$= \int_c^{c+h} \pi y^2 dx, \text{ where } \frac{x^2}{a^2} - \frac{y^2}{b^2} = 1$$

$$= \frac{\pi b^2}{a^2}\int_c^{c+h}(x^2 - a^2)dx = \frac{\pi b^2}{a^2}\left[\frac{x^3}{3} - a^2x\right]_c^{c+h}$$

$$= \frac{\pi b^2}{a^2}\left[\frac{1}{3}\{(c+h)^3 - c^3\} - a^2\{(c+h) - c\}\right]$$

$$= \frac{\pi b^2}{3a^2}[h^3 + 3ch^2 + 3c^2h - 3a^2h] = V_2, \text{ say.}$$

$\therefore$ the required volume $= V_1 - V_2$

$$= \frac{\pi b^2}{3a^2}[h^3 + 3ch^2 + 3c^2h] - \frac{\pi b^2}{3a^2}[h^3 + 3ch^2 + 3c^2h - 3a^2h]$$

$$= \frac{\pi b^2}{3a^2} \cdot 3a^2h$$

$= \pi b^2 h =$ volume of the cylinder of radius b and height h.

Example 6:

Find the volume of the solid generated by the revolution of the curve $y = a^3/(a^2 + x^2)$ about its asymptote.

Solution:

The given curve is $y = a^3/(a^2 + x^2)$

or $$x^2y = a^2(a - y). \quad ...(1)$$

Equating to zero, the coefficient of the highest power of x, the asymptote parallel to x-axis is y = 0 *i.e.*, x-axis. The shape of the curve is as shown in the figure. Take an elementary strip PMNQ where P (x, y) and θ (x + δx,y + δy) are two neighbouring points on the curve. We have PM = y and MN = δx.

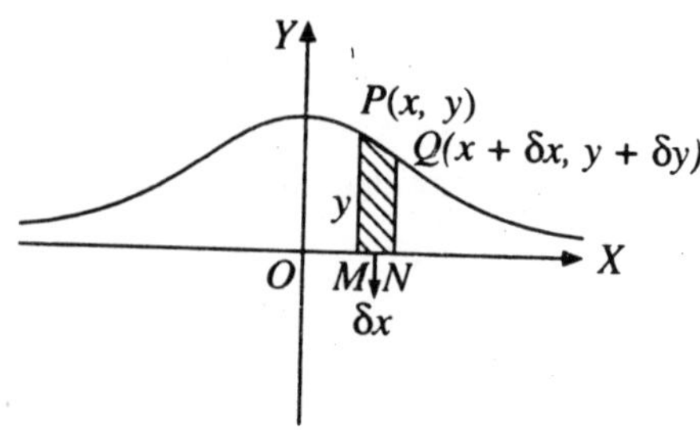

Fig. 3.8

Now volume of the elementary disc formed by revolving the strip PMNQ about the axis of x is $\pi\ PM^2.MN = \pi y^2 \delta x$.

The curve is symmetrical about y-axis and for the portion of the curve in the positive quadrant x varies from 0 to ∞.

$\therefore$ the required volume

$$=2\int_0^{\infty} \pi y^2 dx = 2\pi \int_0^{\infty} \frac{a^6}{(x^2+a^2)^2} dx, \qquad \text{from (1)}$$

$$=2\pi\, a^6 \int_0^{\pi/2} \frac{a \sec^2 \theta d\theta}{a^4 (1+\tan^2 \theta)^2}, \text{ putting } x = a \tan\theta \text{ so that } dx = a \sec^2\theta\, d\theta$$

$$=2\pi\, a^3 \int_0^{\pi/2} \frac{\sec^2 \theta d\theta}{\sec^4 \theta}$$

$$= 2\pi\, a^3 \int_0^{\pi/2} \cos^2 \theta d\theta$$

$$=2\pi\, a^3 \cdot \frac{1}{2} \cdot \frac{1}{2} \pi = \frac{1}{2} \pi^2 a^3.$$

Example 7:

The curve $y^2\ (a + x) = x^2\ (3a - x)$ revolves about the axis of x. Find the volume generated by the loop.

Solution:

The given curve is $y^2\ (a + x) = x^2\ (3a - x)$ ...(1)

It is symmetrical about x-axis. Putting y = 0 in (1), we get x = 0 and x = 3a *i.e.*, a loop is formed between (0, 0) and (3a, 0).

The volume generated by the revolution of the whole loop about x-axis is the same as the volume generated by the revolution of the upper half of the loop about x-axis.

Take an elementary strip PMNQ where P is the point (x, y) and Q is the point (x + δx, y + δy). We have PM = y and MN = δx.

Now volume of the elementary disc formed by revolving the strip PMNQ about he axis of x is = πPM^2. MN = $\pi y^2\, dx$.

∴ the required volume generated by the loop

$$= \int_a^{3a} \pi y^2 dx = \pi \int_0^{3a} \frac{x^2(3a - x)}{a + x} dx,$$

$$= \pi \int_0^{3a} \left\{ -x^2 + 4ax - 4a^2 + \frac{4a^4}{x + a} \right\} dx,$$

dividing the Nr. By the Dr.

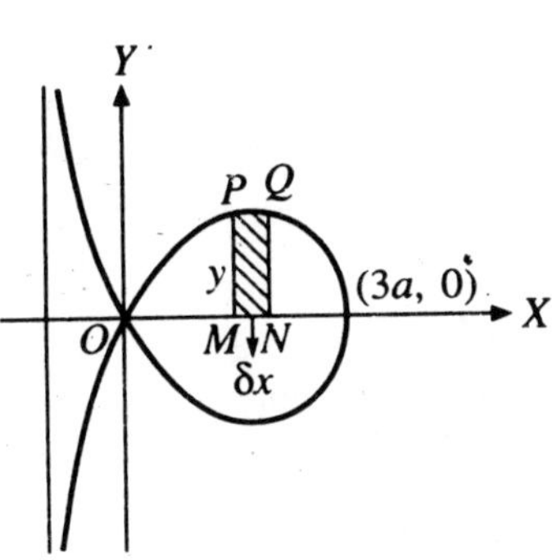

Fig. 3.9

$$= \pi \left[-\frac{x^3}{3} + \frac{4ax^2}{2} - 4a^2x + 4a^3 \log(x + a) \right]_0^{3a}$$

$$= \pi\, [-\, 9a^3 + 18a^3 - 12a^3 + 4a^3 (\log 4a - \log a)]$$

$$= \pi\, [-\, 3a^3 + 4a^3 \log 4] = \pi a^3\, [8 \log 2 - 3].$$

Example 8:

Find the volume formed by the revolution of the loop of the curve $y^2 (a + x) = x^2 (a - x)$ about the axis of x.

Solution:

Do your self.

The required volume = $2\pi a^3 \left(\log 2 - \frac{2}{3} \right)$.

Example 9:

Find the volume of the solid generated by the revolution of the loop of the curve $y^2 = x^2 (a - x)$ about the axis of x.

Solution:

The given equation of the curve is $y^2 = x^2 (a - x)$ ...(1)

Putting y = 0 in (1) we get x = 0, x = a

i.e., the loop is formed between (0, 0) and (a, 0).

∴ the required volume = $\int_0^a \pi y^2 dx$,

$$= \pi\int_0^a x^2 (a - x)dx, \qquad \text{from (1)}$$

$$= \pi\int_0^a (x^2 a - x^3)dx = \pi\left[a\frac{x^3}{3} - \frac{x^4}{4}\right]_0^a$$

$$= \pi a^4\left[\frac{1}{3} - \frac{1}{4}\right] = \frac{1}{12}\pi a^4.$$

Example 10:

Show that volume of the solid generated by the revolution of the upper half of the loop the curve $y^2 = x^2 (2 - x)$ about x-axis is $4/3\pi$.

Solution:

Do your self.

Put $a = 2$.

Example 11:

Find the volume of the spindle shaped solid generated by revolving the astroid $x^{2/3} + y^{2/3} = a^{2/3}$ about the x-axis.

Solution:

The given curve is $x^{2/3} + y^{2/3} = a^{2/3}$. ...(1)

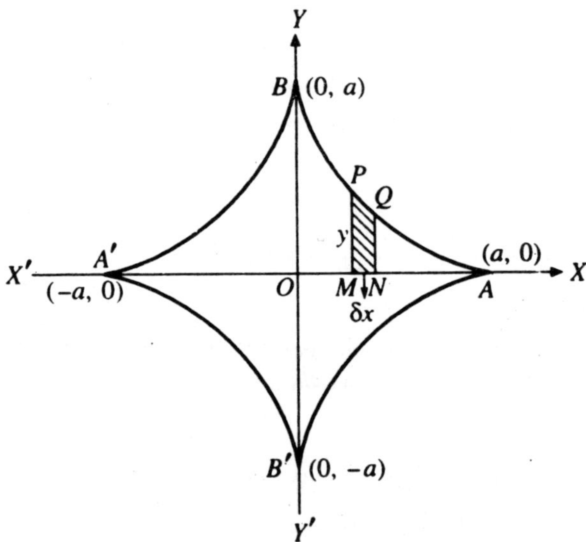

Fig. 3.10

The curve is symmetrical about both the axes. The coordinates of B are (0, a) and those of A are (a, 0).

Take an elementary strip PMNQ where P is the point (x, y) and Q is the point $(x + \delta x, y + \delta y)$ on the curve. We have PM = y and MN = δx.

Now volume of the elementary disc formed by revolving the strip PMNQ about the axis of x is $= \pi y^2 \delta x$.

$$\therefore \quad \text{the required volume} = 2\int_0^a \pi y^2 dx, \text{ by symmetry}$$

$$= 2\pi \int_0^a (a^{2/3} - x^{2/3})^3 dx$$

$$[\because \text{ from (1), } y^{2/3} = (a^{2/3} - x^{2/3}) \text{ so that } y^2 = (a^{2/3} - x^{2/3})^3]$$

$$= 2\pi \int_0^{\pi/2} a^2 \cos^6\theta \, . \, 3a \sin^2\theta \cos\theta \, d\theta,$$

putting $x = a \sin^3\theta$

so that $dx = 3a \sin^2\theta \cos\theta \, d\theta$

$$= 6\pi a^3 \int_0^{\pi/2} \sin^2\theta \cos^7\theta d\theta$$

$$= 6\pi a^3 \cdot \frac{1.6.4.2}{9.7.5.3.1} = \frac{32\pi a^3}{105}.$$

Example 12:

The area of the curve $x^{2/3} + y^{2/3} = a^{2/3}$ lying in the first quadrant revolves about x-axis. Find the volume of the solid generated.

Solution:

Do your self.

$$\text{The required volume} = \frac{1}{2} \cdot \frac{32\pi a^3}{105} = \frac{16}{105}\pi a^3$$

Example 13:

Find the volume of the solid obtained by revolving the loop of the curve $a^2y^2 = x^2 (2a - x) (x - a)$ about x-axis.

Solution:

The given curve is $a^2y^2 = x^2 (2a - x) (x - a)$. ...(1)

The curve (1) is symmetrical about x-axis It passes through the origin but the origin is a conjugate point. The curve cuts the x-axis at the points (a, 0) and (2a, 0). and so the loop of the curve is formed between (a, 0) and (2a, 0).

$\therefore$ the required volume $= \int_{x=a}^{2a} \pi y^2 dx$

$$= \pi \int_a^{2a} \frac{x^2(2a-x)(x-a)}{a^2} dx, \text{ form (1)}$$

$$= \frac{\pi}{a^2} \int_a^{2a} (-x^4 + 3ax^3 - 2a^2x^2)\, dx$$

$$= \frac{\pi}{a^2} \left[-\frac{x^5}{5} + \frac{3ax^4}{4} - \frac{2a^2x^3}{3} \right]_a^{2a}$$

$$= \frac{\pi}{a^2} \left[\left(-\frac{32a^5}{5} + 12a^5 - \frac{16a^5}{3} \right) - \left(\frac{a^5}{5} + \frac{3a^5}{4} - \frac{2a^5}{4} \right) \right]$$

$$= \pi a^3 \left[-\frac{32}{5} + 12 - \frac{16}{3} - \frac{1}{5} - \frac{3}{4} + \frac{2}{3} \right] = \frac{23}{60} \pi a^3.$$

Example 14:

A basin is formed by the revolution of the curve $x^3 = 64y$, $(y > 0)$ about the axis of y. If the depth of the basin is 8 inches, how many cubic inches of water it will hold?

Solution:

The given curve is $x^3 = 64y$. ...(1)

The curve (1) passes through the origin and the tangent there is the line $y = 0$. When $y > 0$, we have $x > 0$ and so no portion of the curve lies in the second quadrant.

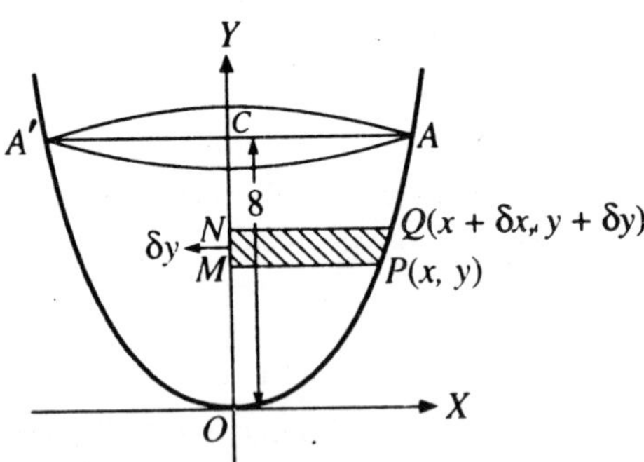

Fig. 3.11

Take an elementary strip PMNQ where P (x, y) and Q (x + δx, y + δy) are two neighbouring points on the curve and PM and QN are perpendiculars from the points P and Q respectively, on the y-axis.

We have PM = x and MN = δy.

Now volume of the elementary disc formed by revolving the strip PMNQ about y-axis is

$$= \pi.\ PM^2.MN = \pi x^2\ \delta y.$$

Clearly to form the required basin y varies from 0 to 8.

∴ the required volume (*i.e.*, the capacity in cubic inches)

$$= \int_{y=0}^{8} \pi x^2 dy = \int_0^8 \pi (64)^{2/3}, \qquad \text{from (1)}$$

$$= 16\pi \int_0^8 y^{2/3} dy = 16\pi \cdot \left[\frac{3}{5} y^{5/3}\right]_0^8 = \frac{48\pi}{5} \cdot 32$$

$$= \frac{1536\pi}{5} \text{ cubic inches.}$$

Revolution about any axis:

Example 15:

Find the volume of the solid generated by the revolution of the cissoid $y^2\ (2a - x) = x^3$ about its asymptote.

Solution:

The given curve is $y^2\ (2a - x) = x^3$. Its shape is as shown in the figure. Equating to zero the coefficient of highest power of y, the asymptote parallel to the axis of y is x = 2a where P is the point (x, y) and Q is the point (x + δx, y + δy).

We havePM = 2a – x and MN = δy.

Now volume of the elementary disc formed by revolving the strip PMNQ about the line x = 2a is

$$= \pi.\ PM^2.\ MN = \pi\ (2a - x)^2\ \delta y.$$

The given curve is symmetrical about x-axis and for the portion of the curve above x-axis y varies from 0 to ∞.

$$\therefore \text{ the required volume } = 2\int_{y=0}^{\infty} \pi (2a - x)^2\, dy. \qquad ...(1)$$

From the given equation of the curve $y^2\ (2a - x) = x^3$ we observe that the value of x cannot be easily found in terms of y. Hence for the sake of integration we change the independent variable from y to x.

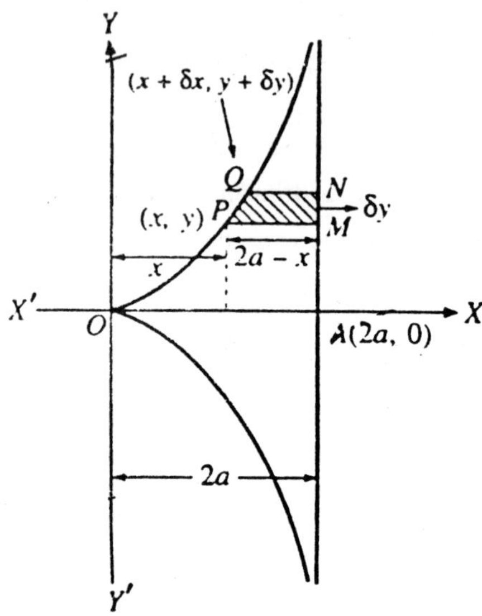

Fig. 3.12

The curve is $y^2 = \dfrac{x^3}{2a - x}$;

$$\therefore \qquad 2y\,\frac{dy}{dx} = \frac{(2a - x)\cdot 3x^2 - x^3(-1)}{(2a - x)^2} = \frac{2(3a - x)x^2}{(2a - x)^2}$$

or
$$dy = \frac{(3a - x)x^2}{(2a - x)^2}\cdot\frac{\sqrt{(2a - x)}}{x\sqrt{x}}\,dx$$

$$= \frac{(3a - x)\sqrt{x}\,\sqrt{(2a - x)}}{(2a - x)^2}\,dx.$$

Also when $y = 0$, $x = 0$

and when $y \to \infty$, $x \to 2a$.

Hence from (1), the required volume

$$= 2\pi\int_{x=0}^{2a}(2a - x)^2\left[\frac{(3a - x)\sqrt{x}\,\sqrt{(2a - x)}}{(2a - x)^2}\right]dx$$

$$= 2\pi\int_{0}^{2a}(3a - x)\sqrt{x}\,\sqrt{(2a - x)}\,dx.$$

Now put $x = 2a \sin^2 \theta$

so that $dx = 4a \sin \theta \cos \theta \, d\theta$.

When $x = 0, \theta = 0$

and when $x = 2a, \theta = \pi/2$.

Therefore the required volume

$$= 2\pi \int_0^{\pi/2} (3a - 2a \sin^2 \theta) \sqrt{(2a)} \sin \theta \sqrt{[2a(1 - \sin^2 \theta)]} \times 4a \sin \theta \cos \theta \, d\theta$$

$$= 16\pi a^3 \int_0^{\pi/2} (3 \sin^2 \theta \cos^2 \theta - 2 \sin^4 \theta \cos^2 \theta) d\theta$$

$$= 16\pi a^3 \left[\frac{3\Gamma\left(\frac{3}{2}\right)\Gamma\left(\frac{3}{2}\right)}{2\Gamma(3)} - \frac{2\Gamma\left(\frac{5}{2}\right)\Gamma\left(\frac{3}{2}\right)}{2\Gamma(4)} \right]$$

$$= 16\pi a^3 \left[\frac{3 \cdot \frac{1}{2} \cdot \sqrt{\pi} \frac{1}{2} \cdot \sqrt{\pi}}{2.2.1} - \frac{2 \cdot \frac{3}{2} \cdot \frac{1}{2} \cdot \sqrt{\pi} \cdot \frac{1}{2} \cdot \sqrt{\pi}}{2.3.2.1} \right]$$

$$= 16\pi a^3 \left[\frac{3\pi}{16} - \frac{\pi}{16} \right] = 2\pi^2 a^3.$$

Note: If the given curve is $y^2 (a - x) = x^3$, then the required volume can be obtained by putting a for 2a in the above Exercise. The volume so obtained is $\frac{1}{2}\pi^2 a^3$.

Remark :

When we are to revolve an area about a line which is neither the x-axis nor the y-axis we take an elementary strip which is perpendicular to the line of revolution as explained in the above example.

Example 16:

The area cut off the parabola $y^2 = 4ax$ by the chord joining the vertex to on end of the latus rectum is rotated through four right angles about the chord. Find the volume of the solid generated.

Solution:

The given parabola is $y^2 = 4ax$.

Let LL' be the latus rectum. The area bounded by the arc OL and the chord OL is revolved about the chord OL.

Let P(x, y) and Q (x + δx, y + δy) be any two neighbouring points on the arc OL and PM, QN be the perpendicuars from P and Q respectively on the axis of revolution OL.

Now volume of the elementary disc formed by revolving the strip PMNQ about the chord OL is

$=\pi.\ PM^2.MN = \pi.\ PM^2.d\ (OM).$

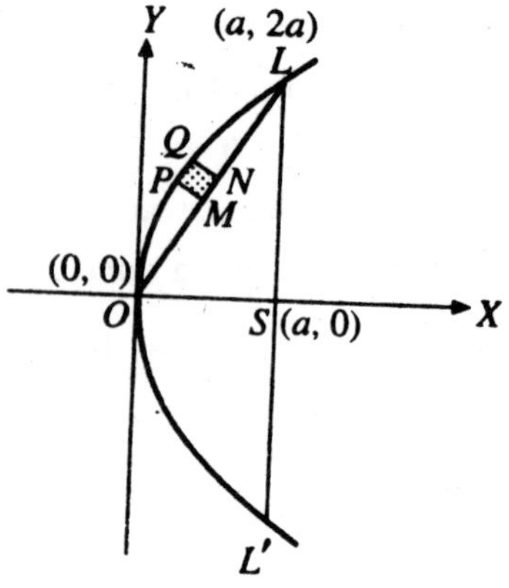

Fig. 3.13

Also equation of the chord OL is

$$y - 0 = \frac{2a-0}{a-0}(x-0)\ i.e.,\ 2x - y = 0 \qquad \text{...(1)}$$

=PM = the length of the perpendicular from (x, y) to (1)

$$= \frac{2x-y}{\sqrt{(2^2+1^2)}} = \frac{2x-y}{\sqrt{5}},$$

and

$$OM = \sqrt{(OP^2 - MP^2)}$$

$$= \sqrt{\left\{(x^2+y^2) - \frac{(2x-y)^2}{5}\right\}} = \frac{x+2y}{\sqrt{5}}.$$

Now the required volume $= \int_{x=0}^{a} \pi\ (PM)^2\ \delta\ (OM).$

[∵ for the arc OL, x varies from 0 to a]

$$= \int_{x=0}^{a} \pi \left(\frac{2x-y}{\sqrt{5}}\right)^2 d\left(\frac{x+2y}{\sqrt{5}}\right)$$

$$= \int_{x=0}^{a} \pi \left(\frac{2x-2\sqrt{(ax)}}{\sqrt{5}}\right)^2 \frac{d}{dx}\left(\frac{x+2.2\sqrt{(ax)}}{\sqrt{5}}\right)dx$$

[∵ $y = 2\sqrt{(ax)}$]

$$= \frac{\pi}{5\sqrt{5}} \int_0^a \{2x - 2\sqrt{(ax)}\}^2 \left(1 + 4\sqrt{a} \cdot \frac{1}{2\sqrt{x}}\right)dx,,$$

$$= \frac{4\pi}{5\sqrt{5}} \int_0^a [x^2 - 2\sqrt{a}x^{3/2} + ax]\left[1 + 2\sqrt{\left(\frac{a}{x}\right)}\right]dx$$

$$= \frac{4\pi}{5\sqrt{5}} \int_0^a [x^2 - 3ax + 2a^{3/2}\sqrt{x}]dx$$

$$= \frac{4\pi}{5\sqrt{5}} \left[\frac{x^3}{3} - \frac{3ax^2}{2} + \frac{2a^{3/2}x^{3/2}}{3/2}\right]_0^a$$

$$= \frac{4\pi}{5\sqrt{5}} \left[\frac{a^3}{3} - \frac{3a^3}{2} + \frac{4a^3}{3}\right] = \frac{2\pi a^3}{15\sqrt{5}} = \frac{2\sqrt{5}}{75}\pi a^2.$$

Example 17:

The area between a parabola and its latus rectum revolves about the directrix. Find the ratio of the volume of the rign thus obtained to the volume of the sphere whose diameter is the latus rectum.

Solution:

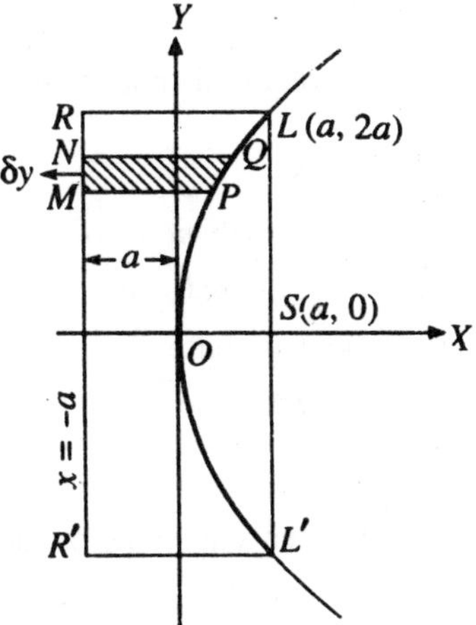

Fig. 3.14

Let the parabola be $y^2 = 4ax$. Then the directrix is the line $x = -a$. Let LL' be the latus rectum. The area LOL'SL is revolved about the directrix. The volume of the ring thus obtained = the volume V_1 of the cylinder formed by the revolution of the rectangle LL'R'R about the directrix – the volume V_2 of the reel formed by the revolution of the arc LOL' about the directrix.

Now the volume V_1 of the cylinder

$= \pi r^2 h = \pi\ (LR)^2.\ LL'$

$= \pi\ (2a)^2.\ 4a = 16\pi a^3.$

To find the volume V_2 of the reel consider an elementary strip PMNQ where P (x, y) and $\sqrt{}$ $(x + \delta x, y + \delta y)$ are two neighbouring points on the arc OL and PM, QN are perpendiculars from P and Q on the directrix.

We have $PM = a + x$ and $MN = \delta y$.

$\therefore$ the volume V_2 of the reel

$$= 2\int_0^{2a} \pi(a + x)^2\, dy, \text{ [by symmetry about x-axis]}$$

$$= 2\int_0^{2a} \pi(a^2 + 2ax + x^2)dy$$

$$= 2\pi\int_0^{2a}\left(a^2 + 2a\cdot\frac{y^2}{4a} + \frac{y^4}{16a^2}\right)dy$$

[$\because$ $x = y^2/4a$]

$$= 2\pi\left[a^2 y + \frac{1}{2}\frac{y^3}{3} + \frac{1}{16a^2}\cdot\frac{y^5}{5}\right]_0^{2a}$$

$$= 2\pi\left[2a^3 + \frac{4}{3}a^3 + \frac{2}{5}a^3\right] = 2\pi a^3\cdot\frac{56}{15} = \frac{112\pi a^3}{15}.$$

$\therefore$ Volume of the ring = volume of the cylinder – volume of the reel

$$= V_1 - V_2 = 16\pi a^3 - \frac{112}{15}\pi a^3 = \frac{128}{15}\pi a^3.$$

Volume of the sphere whose diameter is the latus rectum 4a *i.e.*, the radius is $= \frac{4}{3}\pi r^3 = \frac{4}{3}\pi(2a)^3 = \frac{32}{3}\pi a^3.$

$$\therefore \text{the required ratio} = \frac{128\pi a^3/15}{32\pi a^3/3} = \frac{4}{5}.$$

Example 18:

Find the volume of the solid formed by revolving the cycloid

$x = a(\theta - \sin\theta)$, $y = a(1 - \cos\theta)$

(i) about its base

(ii) *about the y-axis.*

Solution:

The given equations of the cycloid are

$$x = a(\theta - \sin\theta),\ y = a(1 - \cos\theta). \qquad ...(1)$$

(i) The arc OBA is revolved about the base *i.e.*, the x-axis. For the arc OBA, θ varies from 0 to 2π and at B, $\theta = \pi$.

Take an elementary strip PMNQ where P is the point (x, y) and Q is the point $(x + \delta x, y + \delta y)$. We have PM = y and MN = δx.

Now the volume of the elementary disc formed by revolving the strip PMNQ about the base (*i.e.*, the x-axis) is

$\pi\ PM^2.MN = \pi y^2 \delta x.$

Now the cycloid is symmetrical about the line BH.

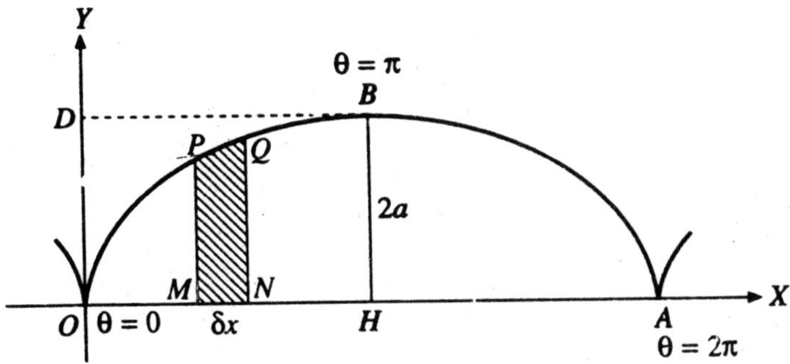

Fig. 3.15

∴ the required volume $= 2\int xy^2 dx$, the limits of integration being extended from O to B

$$= 2\pi\int_{\theta=0}^{\pi} y^2 \frac{dx}{d\theta} d\theta$$

$$= 2\pi\int_0^{\pi} a^2(1-\cos\theta)^2 a(1-\cos\theta)d\theta, \quad \text{form (1)}$$

$$= 2\pi\int_0^{\pi} a^3(1-\cos\theta)^3 d\theta$$

$$= 2\pi a^3\int_0^{\pi}\left(2\sin^2\frac{\theta}{2}\right)^3 d\theta = 16\pi a^3\int_0^{\pi}\sin^6\frac{\theta}{2}d\theta$$

$$= 32\pi a^3\int_0^{\pi/2}\sin^6\phi d\phi \text{ putting } \frac{\theta}{2}$$

$= \phi$ so that $d\theta = 2d\phi$

$$= 32\pi a^3\cdot\frac{5}{6}\cdot\frac{3}{4}\cdot\frac{1}{2}\cdot\frac{1}{2}\pi = 5\pi^2 a^3.$$

(ii) When the curve revolves about y-axis, the required volume of the solid generated = the volume generated by the revolution of the area OABDO about y-axis – the volume generated by the revolution of the area OBDO about the y-axis. ...(2)

Also at A, $\theta = 2\pi$; at B, $\theta = \pi$ and at O, $\theta = 0$.

Now the area OABD is bounded by the arc AB of the cycloid and the axis of y. Therefore volume of the solid generated by the revolution of the area OABDO about y-axis

$$= \int_{\theta=2\pi}^{\pi}\pi x^2 dy = \int_{\theta=2\pi}^{\pi}\pi x^2\frac{dy}{d\theta}d\theta$$

$$= \pi\int_{2\pi}^{\pi} a^2(\theta - \sin\theta)^2 a\sin\theta d\theta, \qquad \text{[from (1)]}$$

$$= \pi\int_{2\pi}^{\pi} a^2(\theta^2 - 2\theta\sin\theta + \sin^2\theta)a\sin\theta d\theta$$

$$= \pi a^3\int_{2\pi}^{\pi} (\theta^2\sin\theta - 2\theta\sin^2\theta + \sin^3\theta)d\theta$$

$$= \pi a^3\int_{2\pi}^{\pi} [\theta^2\sin\theta - \theta(1-\cos 2\theta) + \frac{1}{2}(3\sin\theta - \sin 3\theta)]d\theta$$

$$= \pi a^3\Big[\theta^2\cdot(-\cos\theta) - 2\theta(-\sin\theta) + 2\cos\theta - \frac{1}{2}\theta^2 + \theta\left(\frac{1}{2}\sin 2\theta\right)$$

$$-1\left(-\frac{1}{4}\cos 2\theta\right) - \frac{3}{4}\cos\theta + \frac{1}{12}\cos 3\theta\Big]_{2\pi}^{\pi},$$

the values of the integrals $\int\theta^2\sin\theta\, d\theta$ and $\int\theta\cos 2\theta\, d\theta$ have been written after applying integration by parts

Again volume of the solid generated by the revolution of the area OBDO about y-axis

$$= \int_{\theta=0}^{\pi}\pi x^2 dy = \int_{\theta=0}^{\pi}\pi x^2\frac{dy}{d\theta}d\theta$$

$$= \pi\int_0^a a^2(\theta - \sin\theta)^2\cdot a\sin\theta\, d\theta$$

$$= \pi a^3\int_0^{\pi}(\theta^2 - 2\theta\sin\theta + \sin^2\theta)\sin\theta\, d\theta$$

$$= \pi a^3\int_0^{\pi}(\theta^2\sin\theta - 2\theta\sin^2\theta + \sin^3\theta)d\theta$$

$$= \pi a^3\int_0^{\pi}\left[\theta^2\sin\theta - \theta(1-\cos 2\theta) + \frac{1}{4}(3\sin\theta - \sin 3\theta)\right]d\theta$$

$$= \pi a^3\int_0^{\pi}\Big[\theta^2(-\cos\theta) - 2\theta(-\sin\theta) + 2\cos\theta - \frac{1}{2}\theta^2 + \theta\left(\frac{1}{2}\sin 2\theta\right)$$

$$-1\left(-\frac{1}{4}\cos 2\theta\right) - \frac{3}{4}\cos\theta + \frac{1}{12}\cos 3\theta\Big]_0^{\pi}$$

$$= \pi a^3\left[\left(\pi^2 - 2 - \frac{1}{2}\pi^2 + \frac{1}{4} + \frac{3}{4} - \frac{1}{12}\right) - \left(2 + \frac{1}{4} - \frac{3}{4} + \frac{1}{12}\right)\right]$$

$$= \pi a^3\left(\frac{1}{2}\pi^2 - \frac{8}{3}\right). \qquad ...(4)$$

$\therefore$ from (2), the required volume = (3) – (4)

$$= \pi a^3\left[\frac{13}{2}\pi^2 - \frac{8}{3}\right] - \pi a^3\left[\frac{1}{2}\pi^2 - \frac{8}{3}\right] = \pi a^3[6\pi^2] = 6\pi^3 a^3.$$

Example 19:

Find the volume of the solid generated by the revolution of the cycloid $x = a(\theta + \sin\theta)$, $y = a(1 - \cos\theta)$, $0 \le \theta \le \pi$,

(i) about the x-axis.

(ii) about the base.

Solution:

As above example. The equation of the cycloid are $x = a(\theta + \sin\theta)$, $y = a(1 - \cos\theta)$.

The cycloid is symmetrical about the y-axis. For half of the curve θ varies from 0 to π.

Example 20:

Prove that the volume of the solid formed by the rotation about the line $\theta = 0$ *of the area bounded by the curve* $r = f(\theta)$ *and the lines* $\theta = \theta_1$, $\theta = \theta_2$ *is*

$$\frac{2}{3}\pi\int_{\theta_1}^{\theta_2} r^3 \sin\theta \, d\theta.$$

Solution:

Let OAB be the area bounded by the curve $r = f(\theta)$ and the radii vectors $\theta = \theta_1$ and $\theta = \theta_2$. We have to find the volume formed by the revolution of the area OAB about the initial line OX.

Take any point $P(r, \theta)$ inside the area OAB and take a small element of the area $r\delta\theta\delta r$ at the point P. Drop PM perpendicular from P to the axis of rotation OX. We have

$$PM = OP \sin\theta = r\sin\theta.$$

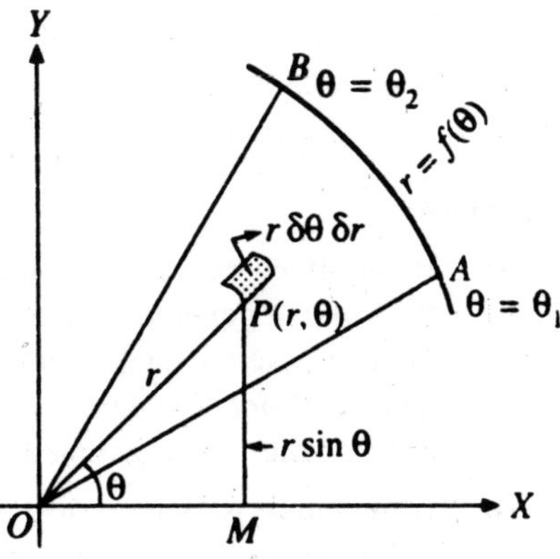

Fig. 3.16

Now the volume of the ring formed by revolving the element of area $r\delta\theta\delta r$ about OX $= 2\pi r\sin\theta . r\delta\theta\delta r = 2\pi r^2 \sin\theta\,\delta\theta\delta r$.

the whole volume formed by revolving the area OAB about OX

$$= \int_{\theta=\theta_1}^{\theta_2} \int_{r=0}^{f(\theta)} 2\pi r^2 \sin\theta \, d\theta \, dr$$

$$= \int_{\theta=\theta_1}^{\theta_2} 2\pi \sin\theta \left[\frac{r^3}{3}\right]_0^{f(\theta)} d\theta$$

$$= \frac{2}{3}\pi \int_{\theta=\theta_1}^{\theta_2} [f(\theta)]^3 \sin\theta d\theta$$

$= \frac{2}{3}\pi \int_{\theta_1}^{\theta_2} r^3 \sin\theta d\theta$, where r is to be replaced from the equation of the curve $r = f(\theta)$.

Note: We can also show that the volume of the solid formed by the rotation of the above mentioned area about the line $\theta = \pi/2$ is equal to

$$\frac{2}{3}\pi \int_{\theta_1}^{\theta_2} r^3 \cos\theta d\theta.$$

The axis of rotation being any line

If, however, the axis of rotation is neither x-axis nor y-axis, but is any other line CD, then *the volume of the solid generated by the revolution about CD of the area bounded by the curve AB, the axis CD and the perpendiculars AC, BD on the aixs is*

$$\int_{OC}^{OD} \pi (PM)^2 d(OM),$$

where PM is the perpendicular drawn from any point P on the curve to the axis of rotation and O is some fixed point on the axis of rotation.

Remarks:

(i) If the given curve is symmetrical about x-axis and we have to find the volume generated by the revolution of the area about x-axis, then in such case we shall revolve only one of the two symmetrical areas and *shall not double it* as in the case of area or length. Obviously each of the two symmetrical parts will generate the same volume.

(ii) If the curve is symmetrical about x-axis and it is required to find the volume generated by the revolution of the area about y-axis, then the volume generated *will be twice* the volume generated by half of the symmetrical portion of the curve.

Example 21:

Show that the volume of a sphere of radius a is $\frac{4}{3}\pi a^3$.

Solution:

The sphere is generated by the revolution of a semi-circular area about its bounding diameter. The equation of the generating circle of radius a and centre as origin is $x^2 + y^2 = a^2$.

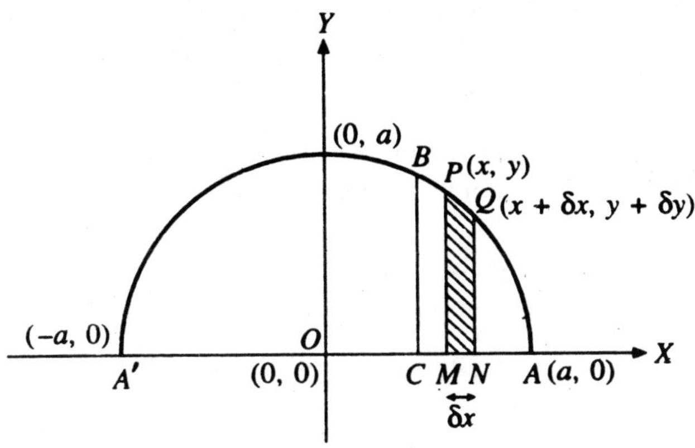

Fig. 3.17

Let AA' be the bounding diameter about which the semi-circle revolves.

Take an elementary strip PMNQ where P is the point (x, y) and Q is the point $(x + \delta x, y + \delta y)$. We have PM = y and MN = δx.

Now volume of the elementary disc formed by revolving the strip PMNQ about the diameter AA' is

$$= \pi . PM^2 . MN = \pi y^2 \delta x = \pi (a^2 - x^2) \delta x.$$

Also the semi-circle is symmetrical about the y-axis and the for the portion of the curve lying in the first quadrant x varies from 0 to a.

$\therefore$ the required volume of the sphere

$$= 2\int_0^a \pi(a^2 - x^2)dx = 2\pi\left[a^2x - \frac{1}{3}x^3\right]_0^a$$

$$= 2\pi\left[a^3 - \frac{1}{3}a^3\right] = \frac{4}{3}\pi a^3$$

Example 22:

Find the volume of a hemisphere.

Solution:

As above example. The hemi-sphere is generated by the revolution of a quadrant of the circle $x^2 + y^2 = a^2$ about x-axis.

The limits for the volume will be from 0 to a. The required volume of hemisphere is $\frac{2}{3}\pi a^3$.

Example 23:

Find the volume of a spherical cap of height h cut off from a sphere of radius a.

Solution:

The limits for the volume of the spherical cap of height h will be from a – h to a. We get the required volume

$$= \int_{a-h}^{a} \pi y^2 dx = \pi\int_{a-h}^{a} (a^2 - x^2)dx = \pi\left[a^2 x - \frac{x^3}{3}\right]_{a-h}^{a}$$

$$= \pi\left[a^3 - \frac{1}{3}a^3 - a^2(a-h) + \frac{1}{3}(a-h)^3\right]$$

$$= \pi\left[a^3 - \frac{1}{3}a^3 - a^3 + a^2 h + \frac{1}{3}a^3 - a^2 h + ah^2 - \frac{1}{3}h^3\right]$$

$$= \pi\left[ah^2 - \frac{1}{3}h^3\right] = \pi h^2\left[a - \frac{1}{3}h\right].$$

Example 24:

A segment is cut off from a sphere of radius a by a plane at a distance 1/2a from the centre. Show that the volume of the segment is 5/32 of the volume of the sphere.

Solution:

Let BC be the line $x = \frac{1}{2}a$.

The segment of the sphere is generated by revolving the area ABCA of the circle about the x-axis. Hence, the limits for the volume of the segment will be from x = 1/2a to x = a.

∴ the volume of the segment of the sphere

$$= \int_{a/2}^{a} \pi y^2 dx = \int_{a/2}^{a} \pi(a^2 - x^2)dx, \qquad [\because x^2 + y^2 = a^2]$$

$$= \pi\left[a^2 x - \frac{x^3}{3}\right]_{a/2}^{a}$$

$$= \pi\left[a^3 - \frac{1}{3}a^3 - \left(\frac{1}{2}a^3 - \frac{1}{24}a^3\right)\right]$$

$$= \pi\left[\frac{5}{24}a^3\right] = \frac{5}{32}\left[\frac{4}{3}\pi a^3\right].$$

Also volume of the sphere $= \frac{4}{3}\pi a^3$.

$$\therefore \quad \frac{\text{Volume of the segment}}{\text{Volume of the sphere}} = \frac{\frac{5}{32}\cdot\left[\frac{4}{3}\pi a^3\right]}{\frac{4}{3}\pi a^3} = \frac{5}{32}.$$

i.e., Volume of the segment = 5/32 of the volume of the sphere.

Example 25:

Find the volume of the paraboloid generated by the revolution about the x-axis of the parabola $y^2 = 4ax$ from $x = 0$ to $x = h$.

Solution:

The given parabola is $y^2 = 4ax$.

Take an elementary strip PMNQ, where P is the point (x, y) and Q is the point $(x + \delta x, y + \delta y)$. Then PM = y and MN = ON – OM = $(x + \delta x) - x = \delta x$. Now volume of the elementary disc formed by revolving the strip PMNQ about the x-axis

$$= \pi.\ PM^2.\ MN = \pi y^2 \delta x.$$

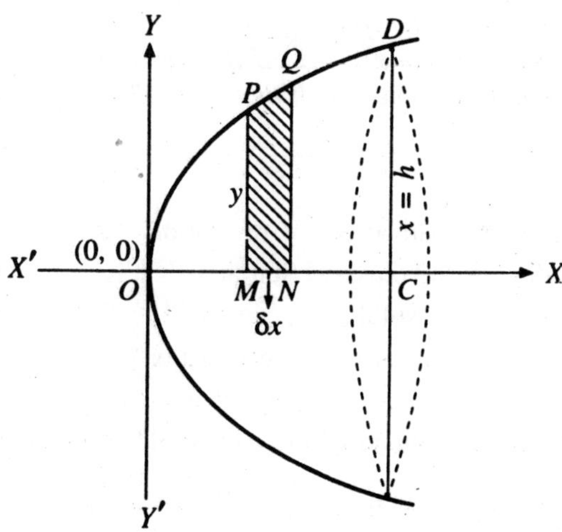

Fig. 3.18

The paraboloid is formed by the revolution of the area ODC about x-axis

Also for the area ODC, x varies from x = 0 to x = h.

∴ the required volume

$$= \int_0^h \pi y^2 dx = \int_0^h \pi(4ax)dx = 4\pi a\left[\frac{x^2}{2}\right]_0^h = 2\pi ah^2.$$

Example 26:

The area of the parabola $y^2 = 4ax$ lying between the vertex and the latus rectum is revolved about the x-axis. Find the volume generated.

Solution:

At the vertex we have x = 0 and for the latus rectum we have x = a. Therefore the limits for the volume generated by revolving the area of the parabola $y^2 = 4ax$ between the vertex and the latus rectum are from x = 0 to x = a.

∴ the required volume

$$= \int_0^a \pi y^2 dx = \pi\int_0^a 4ax\, dx, \qquad [\because y^2 = 4ax]$$

$$= 4a\pi\left[\frac{1}{2}x^2\right]_0^a = 4a\pi \cdot \frac{1}{2}a^2 = 2\pi a^3.$$

Example 27:

A paraboloid of revolution is generated by rotating the parabola $y^2 = 4ax$ about OX. Find the volume generated by that portion of the curve which lies between x = 0 and x = L. If R is the area of the cross-section at x = L, show that the volume is half that of a cylinder of base area R and length L.

Solution:

The volume of the paraboloid of revolution $= 2\pi aL^2$. ...(1)

Also the radius of the cross-section at x = L is

$$\sqrt{(4aL)}, \qquad (\because y^2 = 4ax)$$

$$\therefore \quad R = \text{area of cross-section at } x = L$$

$$= \pi \cdot (\text{radius})^2 = \pi\,\{\sqrt{(4aL)}\}^2 = 4a\pi L.$$

Now from (1), the volume of the paraboloid of revolution

$$= 2\pi aL^2 = \frac{1}{2}\cdot(4\pi aL^2)$$

$$= \frac{1}{2}[4a\pi L \times L]$$

$$= \frac{1}{2}[R \times L]$$

$=\frac{1}{2}$ volume of the cylinder of base area R and length L.

Example 28:

The part of the parabola $y^2 = 4ax$ cut off by the latus rectum revolves about the tangent at the vertex. Find the volume of the reel thus generated.

Solution:

The given parabola is $y^2 = 4ax$. It is symmetrical about x-axis. The tangent at the vertex is $x = 0$ *i.e.*, y-axis. LL' is the latus rectum.

A reel is formed by revolving about y-axis the area enclosed between the arc L'OL of the parabola and the axis of y.

The volume of the reel generated by the revolution of the arc cut off by the latus rectum LL' about y-axis = 2 × volume generated by revolving the area OLK about y-axis. Consider an elementary strip PMNQ parallel to the axis of x, where P is the axis of x, where P is the point (x, y) and Q is the point $(x + \delta x, y + \delta y)$ on the parabola $y^2 = 4ax$. Then PM = x and $NM = ON - OM = (y + \delta y) - y = \delta y$.

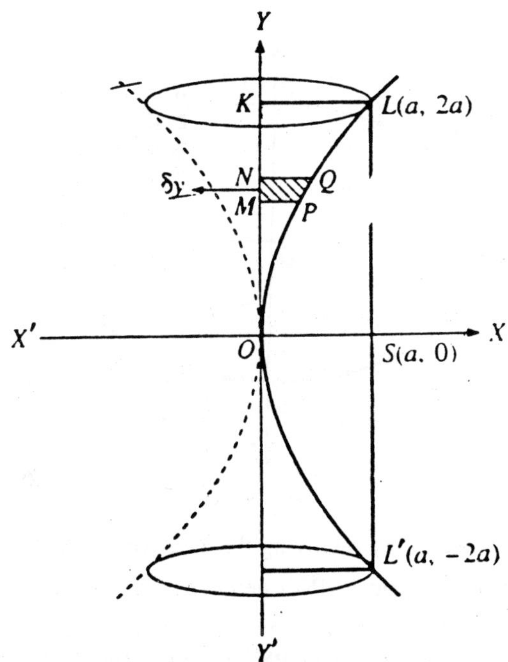

Fig. 3.19

Now volume of the elementary disc formed by revolving the strip PMNQ about y-axis = π $(PM)^2 \cdot (NM) = \pi x^2 \delta y$.

Also as the length of the semi-latus rectum SL is 2a, therefore varies from 0 to 2a.

$\therefore$ the required volume

$$= 2\int_0^{2a} \pi x^2 dy = 2\int_0^{2a} \pi \left[\frac{y^2}{4a}\right]^2 dy, \quad (\because y^2 = 4ax)$$

$$= \frac{\pi}{8a^2}\int_0^{2a} y^4 dy = \frac{\pi}{8a^2}\left[\frac{y^5}{5}\right]_0^{2a}$$

$$= \frac{\pi}{40a^2} \cdot 32a^5 = \frac{4}{5}\pi a^3.$$

Example 29:

Find the volume of the solid generated by revolving the ellipse $x^2/a^2 + y^2/b^2 = 1$ about the x-axis.

Solution: -

The given equation of the ellipse is $x^2/a^2 + y^2/b^2 = 1$ ...(1)

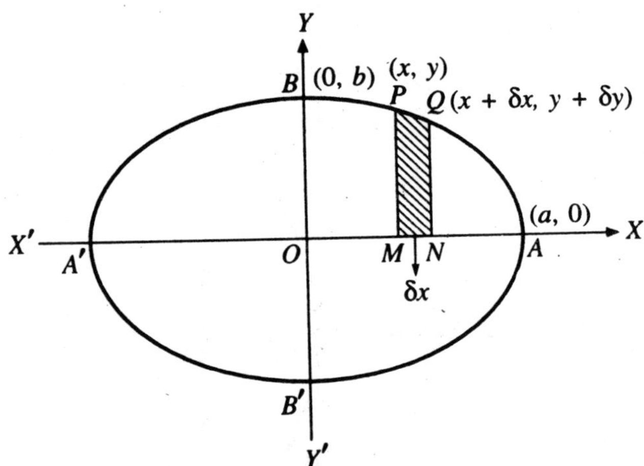

Fig. 3.20

The solid is generated by revolving the area ABA'OA about the x-axis.

Take an elementary strip PMNQ perpendicular to the axis of x.

We have PM = y and MN = δx. Now volume of the elementary disc formed by revolving the strip PMNQ about the x-axis = $\pi . (PM)^2 . MN = \pi y^2 \delta x$.

Also the ellipse is symmetrical about the y-axis and for the portion of the curve lying in the first quadrant x varies from 0 to a. Therefore the required volume of the solid formed

$$= 2\pi\int_0^a y^2 dx = 2\pi\int_0^a \frac{b^2}{b^2}(a^2 - x^2)\,dx, \text{ from (1)}$$

$$= 2\pi\,\frac{b^2}{a^2}\left[a^2x - \frac{1}{3}x^3\right]_0^a$$

$$= 2\pi\,\frac{b^2}{a^2}\left(a^3 - \frac{1}{3}a^3\right) = \frac{4}{3}\pi ab^2.$$

Example 30:

The part of the ellipse $x^2/a^2 + y^2/b^2 = 1$ cut off by a latus rectum revolves about the tangent at the nearer vertex. Find the volume of the reel thus generated.

Solution:

The given ellipse is $x^2/a^2 + y^2/b^2 = 1$. Let LSL' be a latus rectum.

The tangent at the nearer vertex A is the line x = a.

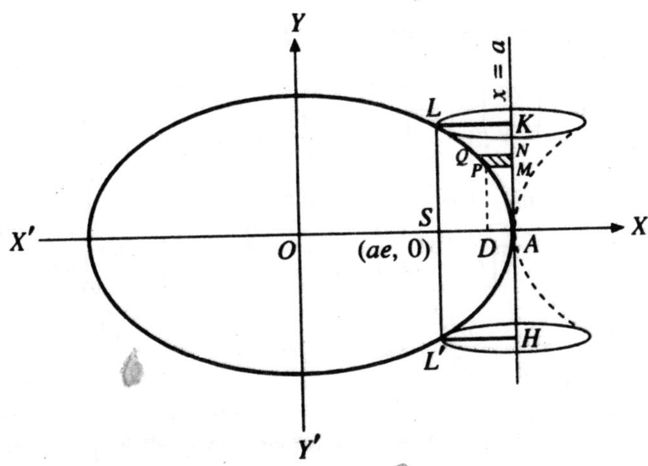

Fig. 3.21

Draw LK and L'H perpendicular to the line x = a.

We have to find volume of the reel generated by the revolution of the area LAL'HKL about the line x = a. Take an elementary strip PMNQ perpendicular to the line x = a, where P (x, y) and Q (x + δx, y + δy) are two neighborliness points on the arc AL of the ellipse and PM and QN are perpendiculars from P and Q respectively to the tangent at A.

Then $\qquad PM = OA - OD = a - x$

and $\qquad MN = AN - AM = y + \delta y - y = \delta y.$

Now volume of the elementary disc formed by revolving the strip PMNQ about the line $x = a$

$$= \pi\,(PM)^2.\ MN = \pi\,(a - x)^2\,\delta y.$$

Also as the length of semi-latus rectum SL is (b^2/a), therefore on the ellipse from A to L, y varies from $y = 0$ to $y = b^2/a$.

Now the required volume of the reel thus generated $= 2 \times$ volume generated by revolving the area LAKL about the tangent at the vertex A

$$=2\int_0^{b^2/a} \pi(a-x)^2\,dy, \qquad \text{where } \frac{x^2}{a^2}+\frac{y^2}{b^2}=1$$

$$=2\pi\int_0^{b^2/a} (a^2-2ax+x^2)\,dy, \qquad \text{where } x^2=\frac{a^2}{b^2}(b^2-y^2)$$

$$=2\pi\int_0^{b^2/a}\left\{a^2-2a\cdot\frac{a}{b}\sqrt{(b^2-y^2)}+\frac{a^2}{b^2}(b^2-y^2\right\}dy$$

$$=\frac{2\pi a^2}{b^2}\int_0^{b^2/a}\{2b^2-2b\sqrt{(b^2-y^2)}-y^2\}\,dy$$

$$=\frac{2\pi a^2}{b^2}\left[2b^2y-2b\left\{\frac{1}{2}y\sqrt{(b^2-y^2)}+\frac{1}{2}b^2\sin^{-1}\left(\frac{y}{b}\right)\right\}-\frac{y^3}{3}\right]_0^{b^2/a}$$

$$=\frac{2\pi a^2}{b^2}\left[2b^2\cdot\frac{b^2}{a}-2b\left\{\frac{1}{2}\frac{b^2}{a}\cdot\sqrt{\left(b^2-\frac{b^4}{a^2}\right)}+\frac{1}{2}b^2\sin^{-1}\frac{b}{a}\right\}-\frac{b^6}{3a^3}\right]$$

$$=\frac{2\pi a^2}{b^2}\cdot\left[2\frac{b^4}{a}-\frac{b^4}{a^2}\cdot\sqrt{(a^2-b^2)}-b^3\sin^{-1}\frac{b}{a}-\frac{b^6}{3a^3}\right]$$

$$=\frac{2\pi a^2}{b^2}\left\{6a^2b-3ab\sqrt{(a^2-b^2)}-3a^3\sin^{-1}\frac{b}{a}-b^3\right\}.$$

Example 31:

Show that the volume of the solid generated by the revolution of the cycloid $x = a(\theta + \sin\theta)$, $y = a(1 - \cos\theta)$, $0 \le \theta \le \pi$, about the y-axis is $\pi a^3\left(\frac{3}{2}\pi^2-\frac{8}{3}\right)$.

Solution:

The curve is as shown in the figure.

The required volume

$$= \int_{y=0}^{2a} \pi x^2 dy = \pi \int_{\theta=0}^{\pi} x^2 \frac{dy}{d\theta} d\theta$$

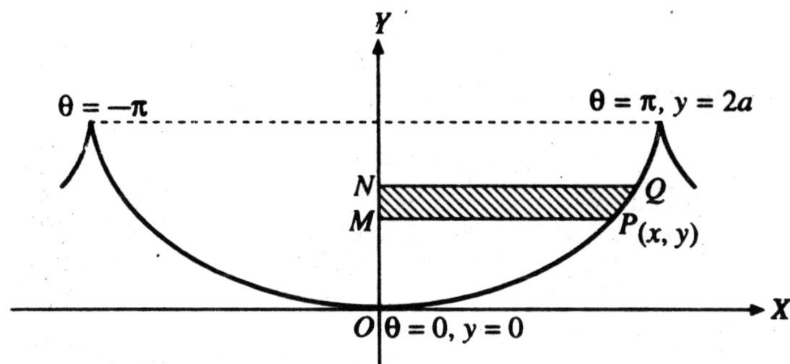

Fig. 3.22

$$= \pi \int_{\theta=0}^{\pi} a^2 (\theta + \sin\theta)^2 a \sin\theta \, d\theta$$

$$= \pi a^3 \int_{\theta=0}^{\pi} (\theta^2 + 2\theta \sin\theta + \sin^2\theta) \sin\theta d\theta$$

$$= \pi a^3 \int_0^{\pi} (\theta^2 \sin\theta + 2\theta \sin^2\theta + \sin^3\theta) d\theta = \pi a^3 (I_1 + 2I_2 I_3), \text{ say}$$

where $I_1 = \int_0^{\pi} \theta^2 \sin\theta d\theta = \left[-\theta^2 \cos\theta\right]_0^{\pi} + \int_0^{\pi} 2\theta \cos\theta \, d\theta$

$$= \pi^2 + \left[2\theta \sin\theta\right]_0^{\pi} - 2\int_0^{\pi} \sin\theta \, d\theta$$

$$= \pi^2 + 2(-1 - 1) = \pi^2 - 4,$$

$$I_2 = \pi^2 \int_0^{\pi} \theta \sin^2\theta \, d\theta$$

$$= \int_0^{\pi} (\pi - \theta) \sin^2(\pi - \theta) d\theta,$$

$$\left[\int_0^a f(x)dx = \int_0^a f(a - x)dx\right]$$

$$= \int_0^{\pi} \pi \sin^2\theta \, d\theta - \int_0^{\pi} \theta \sin^2\theta \, d\theta$$

$$= \pi \int_0^{\pi} \sin^2 \theta \, d\theta - I_2$$

so that $$2I_2 = \pi \int_0^{\pi} \sin^2 \theta \, d\theta = 2\pi \int_0^{\pi/2} \sin^2 \theta \, d\theta$$

or $$I_2 = \pi \int_0^{\pi/2} \sin^2 \theta \, d\theta = \pi \cdot \frac{1}{2} \cdot \frac{1}{2} \pi = \frac{1}{4} \pi^3,$$

and $$I_3 = \int_0^{\pi} \sin^3 \theta \, d\theta = 2 \int_0^{\pi/2} \sin^3 \theta \, d\theta = 2 \cdot \frac{2}{3.1} = \frac{4}{3}.$$

$\therefore$ the required volume $= \pi a^3 (I_1 + 2I_2 + I_3)$

$$= \pi a^3 \left[(\pi^2 - 4) + 2 \cdot \frac{1}{4} \pi^2 + \frac{4}{3} \right]$$

$$= \pi a^3 \left[\frac{3}{2} \pi^2 - \frac{8}{3} \right].$$

Example 32:

Prove that the volume of the reel formed by the revolution of the cycloid $x = a(\theta + \sin \theta)$, $y = a(1 - \cos \theta)$ about the tangent at the vertex is $\pi^2 a^3$.

Solution:

The given cycloid is symmetrical about the y-axis and the tangent at the vertex is x-axis. The reel is formed by the revolution about x-axis of the area enclosed between the cycloid and the x-axis.

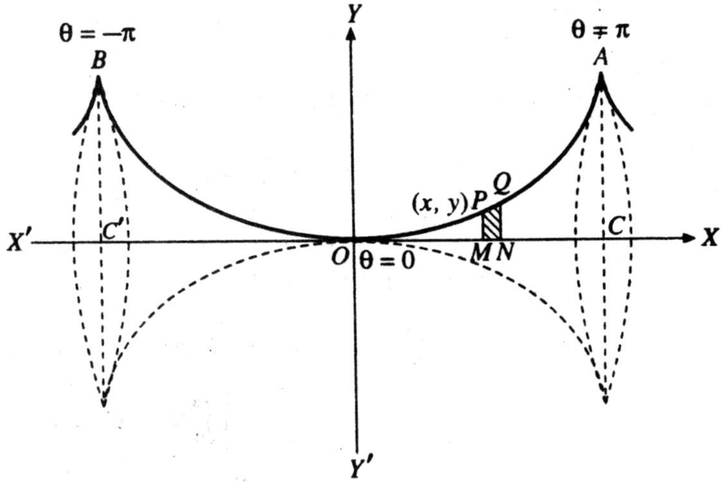

Fig. 3.23

For the arc OA of the curve θ varies from 0 to π.

Take an elementary strip PMNQ where P is the point (x, y) and Q is the point (x + δx, y + δy). We have PM = y and MN = δx.

Now volume of the elementary disc formed by revolving the strip PMNQ about the tangent at the vertex (*i.e.*, about x-axis) is

$$= \pi\ PM^2.MN = \pi y^2\,\delta x.$$

∴ the required volume

$= 2\int \pi y^2 dx$, between the limits of integration from O to A

$$= 2\int_0^{\pi} \pi y^2 \frac{dx}{d\theta} d\theta$$

$= 2\pi\int_0^{\pi} a^2(1-\cos\theta)^2 \cdot a(1+\cos\theta)d\theta$, putting for y and $\frac{dx}{d\theta}$

$$=2\pi a^3\int_0^{\pi}\left(2\sin^2\frac{\theta}{2}\right)^2\cdot\left(2\cos^2\frac{\theta}{2}\right)d\theta$$

$=2\pi a^3\int_0^{\pi/2} 4\sin^4 t.2\cos^2 t.2dt$, putting $\frac{\theta}{2} = t$ so that $d\theta = 2\ dt$

$$=32\ \pi a^3\int_0^{\pi/2}\sin^4 t\cos^2 t dt = 32\ \pi a^3\cdot\frac{3.1.1}{6.4.2}\cdot\frac{\pi}{2} = \pi^2 a^3.$$

Example 33:

Prove that the volume of the solid generated by the revolution about the x-axis of the loop of the curve

$$x = t^2,\ y = t - \frac{1}{3}t^3 \text{ is } \frac{3}{4}\pi$$

Solution:

The given parametric equations of the curve are

$$x = t^2,\ y = t - \frac{1}{3}t^3. \qquad ...(1)$$

Eliminating t, we have

$$y^2 = t^2\left(1-\frac{1}{3}t^2\right)^2 = x\left(1-\frac{1}{3}x\right)^2.$$

The curve is thus symmetrical about the x-axis. The curve cuts the x-axis at the points (0, 0) and (3, 0). Therefore, the loop of the curve lies between these points. Putting y = 0 in (1), we get

$$t\left(1-\frac{1}{3}t^2\right) = 0 \text{ giving } t = 0, \pm\sqrt{3}.$$

Therefore for the upper half of the loop t varies from 0 to √3.

∴ the required volume

$$= \int_0^{\sqrt{3}} \pi y^2 \cdot \frac{dx}{dt}\, dt$$

$$= \int_0^{\sqrt{3}} \pi \left(t - \frac{1}{3}t^3\right)^2 2t\, dt, \text{ from (1)}$$

$$= 2\pi \int_0^{\sqrt{3}} t\left(t^2 + \frac{1}{9}t^6 - \frac{2}{3}t^4\right) dt$$

$$= 2\pi \int_0^{\sqrt{3}} \left(t^3 + \frac{1}{9}t^7 - \frac{2}{3}t^5\right) dt$$

$$= 2\pi \left[\frac{t^4}{4} + \frac{1}{9}\cdot\frac{t^8}{8} - \frac{2}{3}\cdot\frac{t^6}{6}\right]_0^{\sqrt{3}}$$

$$= 2\pi \left[\frac{9}{4} + \frac{1}{9}\cdot\frac{91}{8} - \frac{2}{3}\cdot\frac{27}{6}\right]$$

$$= 2\pi \left[\frac{9}{4} + \frac{9}{8} - 3\right] = \frac{3\pi}{4}.$$

Example 34:

Find the volume of the spindle shaped solid generated by revolving the astroid $x^{2/3} + y^{2/3} = a^{2/3}$ about the x-axis.

Solution:

The parametric equations of the given curve $x^{2/3} + y^{2/3} = a^{2/3}$ are $x = a\cos^3 t$, $y = a\sin^3 t$. ...(1)

The curve is symmetrical about both the axes.

At the point B, x = 0 and so t = 1/2π.

Again at the point A, x = a and so t = 0.

Therefore, for the portion of the curve lying in the first quadrant t varies from 1/2π to 0.

∴ the required volume = 2 × volume generated by revolving the

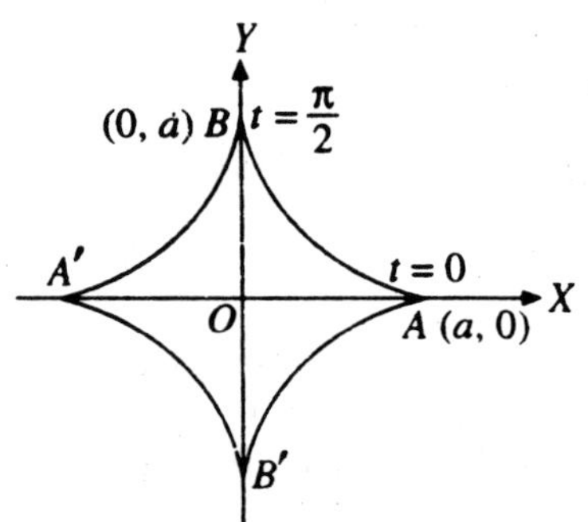

Fig. 3.24

area lying in the Ist quadrant

$$= 2\int_{x=0}^{a} \pi y^2 dx$$

$$= 2\int_{\pi/2}^{0} \pi y^2 \cdot \frac{dx}{dt} dt$$

$$= 2\pi\int_{\pi/2}^{0} a^2 \sin^6 t \cdot (-3a\cos^2 t \sin t\, dt),$$

$$= 6\pi a^3 \int_0^{\pi/2} \sin^7 t \cdot \cos^2 dt$$

$$= 6\pi a^3 \cdot \frac{6.4.2.1}{9.7.5.3.1} = \frac{32}{105}\pi a^3.$$

Example 35:

Find the volume of the solid generated by the revolution of the tractrix $x = a\cos t + 1/2\, a \log \tan^2 (t/2)$, $y = a \sin t$ about its asymptote.

Solution:

The given curve is $x = a\cos t + 1/2\, a \log \tan^2 (t/2)$, $y = a \sin t$. ...(1)

$$\therefore \quad \frac{dx}{dt} = -a\sin t + \frac{1}{2}a \cdot \frac{1}{\tan^2(t/2)} \cdot 2\tan(t/2)\sec^2(t/2) \cdot \frac{1}{2}$$

$$= -a\sin t + \frac{a}{2\sin(t/2)\cos(t/2)}$$

$$= -a\sin t + \frac{a}{\sin t}$$

$$= a\frac{(1-\sin^2 t)}{\sin t} = a\frac{\cos^2 t}{\sin t} \quad ...(2)$$

Now the given curve is symmetrical both the axes and the asymptote is the line $y = 0$ *i.e.*, x-axis.

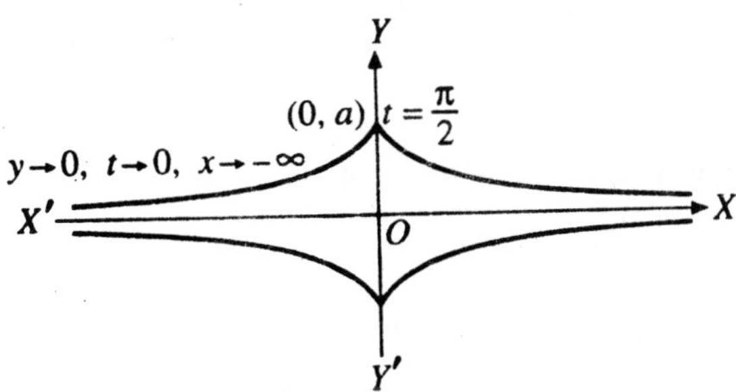

Fig. 3.25

For the portion of the curve lying in the second quadrant y varies from a to 0, t varies from $\pi/2$ to 0 and x varies from 0 to $-\infty$.

$\therefore$ the required volume

$$= 2\int_{-\infty}^{0} \pi y^2 dx$$

$$= 2\int_{0}^{\pi/2} \pi y^2 \frac{dx}{dt}.dt$$

$$= 2\pi\int_{0}^{\pi/2} a^2 \sin^2 t.\frac{a\cos^2 t}{\sin t}dt, \text{ from (1) and (2)}$$

$$= 2\pi a^2\int_{0}^{\pi/2} \cos^2 t \sin t\, dt$$

$$= 2\pi a^3 \frac{1}{3.1} = \frac{2}{3}\pi a^3.$$

Example 36:

Prove that the volume of the solid generated by the revolution of the tractrix $x = a\cos t + \frac{1}{2} a \log \tan^2 \frac{1}{2} t$, $y = a\sin t$, *about its asymptote equals half of a sphere of radius a.*

Solution:

As above example. We get the volume of the solid generated by the revolution of the given tractrix $= \frac{2}{3}\pi a^3 = \frac{1}{2}\cdot\frac{4}{3}\pi a^3 = \frac{1}{2}\cdot$ volume of the sphere of radius a.

Example 37:

Find the volume of the solid generated by the revolution of the cissoid $x = 2\sin^2 t$, $y = 2a\sin^3 t/\cos t$ *about its asymptote.*

Solution:

The given parametric equation of the cissoid are $x = 2a\sin^2 t$, $y = 2a\sin^3 t/\cos t$.

Let us eliminate t between these equation.

We have $\sin^2 t = x/2a$ ...(1)

Now
$$y^2 = \left[2a\frac{\sin^3 t}{\cos t}\right]^2$$

$$= 4a^2 \frac{\sin^6 t}{\cos^2 t}$$

$$= 4a^2 \frac{(\sin^2 t)^3}{1 - \sin^2 t}$$

$$= \frac{\{4a^2 (x/2a)^3\}}{\{1 - (x/2a)\}}, \qquad \text{from (1)}$$

$$= \frac{x^3}{2a - x}.$$

Thus, $y^2 (2a - x) = x^3$ is the cartesian equation of the given cissoid and the for the shape of the curve.

The required volume

$$= 2\pi \int_{y=0}^{\infty} (2a - x)^2 \, dy = 2\pi \int_{t=0}^{\pi/2} (2a - x)^2 \frac{dy}{dt} dt$$

$$\left[\because t = 0, \text{when } y = 0 \text{and } t \to \frac{1}{2}\pi \text{ when } y \to \infty\right]$$

$$= 2\pi \int_0^{\pi/2} (2a - 2a \sin^2 t)^2 \cdot 2a \frac{3 \sin^2 t \cos^2 t + \sin^4 t}{\cos^2 t}$$

$$= 16\pi a^3 \int_0^{\pi/2} \cos^2 t (3 \sin^2 t \cos^2 t + \sin^4 t) dt$$

$$= 16\pi a^3 \left[\int_0^{\pi/2} 3 \sin^2 t \cos^4 t \, dt + \int_0^{\pi/2} \sin^4 t \cos^2 t \, dt\right]$$

$$= 16\pi a^3 \left[3 \cdot \frac{1.3.1}{6.4.2} \cdot \frac{1}{2}\pi + \frac{3.1.1}{6.4.2} \cdot \frac{1}{2}\pi\right]$$

$$= 16\pi a^3 \cdot \frac{\pi}{32} \cdot (3 + 1) = 2\pi^2 a^3.$$

Example 38:

Find the volume of the solid generated by the revolution of $r = 2a \cos \theta$ about the initial line.

Solution:

The given curve $r = 2a \cos \theta$ is a circle passing through the pole. It is symmetrical about the initial line (*i.e.*, x-axis). We have $\theta = 0$ at the point A and $\theta = \pi/2$ at the point O where $r = 0$.

Thus for the upper half of the circle θ varies from 0 to $\frac{1}{2}\pi$.

$\therefore$ the required volume $= \dfrac{2}{3} \displaystyle\int_0^{\pi/2} \pi r^3 \sin \theta \, d\theta$

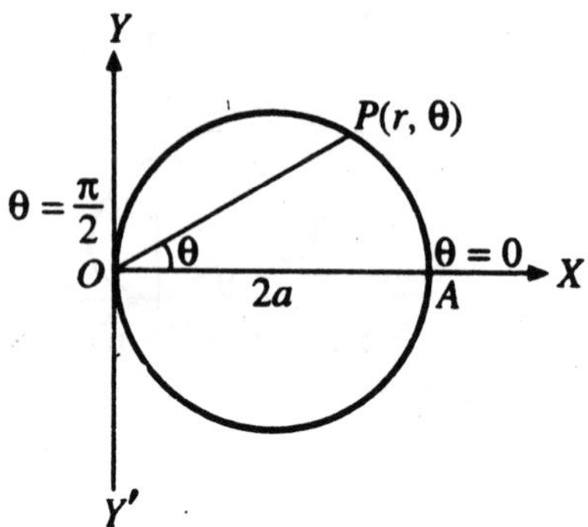

Fig. 3.26

$$= \frac{2}{3}\pi \int_0^{\pi/2} (2a \cos\theta)^3 \sin\theta \, d\theta \qquad [\because r = 2a \cos\theta]$$

$$= \frac{16\pi a^3}{3} \int_0^{\pi/2} \cos^3\theta \sin\theta \, d\theta$$

$$= -\frac{16\pi a^3}{3} \int_0^{\pi/2} \cos^3\theta \cdot (-\sin\theta) d\theta$$

$$= -\frac{16\pi a^3}{3} \cdot \left[\frac{\cos^4\theta}{4}\right]_0^{\pi/2}$$

$$= -\frac{4}{3}\pi a^3 [0 - 1] = \frac{4}{3}\pi a^3.$$

Example 39:

The cardioid r = a (1 + cos θ) revolves about the initial line. Find the volume of the solid thus generated.

Solution:

The given curve is $r = a(1 + \cos\theta)$. ...(1)

It is symmetrical about the initial line.

We have $r = 0$

when $\cos\theta = -1$ *i.e.*, $\theta = \pi$.

Also r is maximum when $\cos\theta = 1$ *i.e.*, $\theta = 0$ and then $r = 2a$. As θ increases from 0 to π, r decreases from 2a to 0.

Hence the shape of the curve is as shown in the figure. For the upper half of the curve, θ varies from 0 to π.

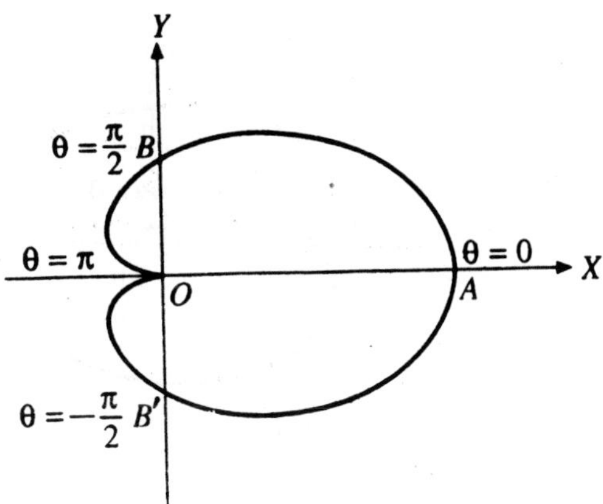

Fig. 3.27

∴ the required volume

$$= \frac{2}{3}\int_0^{\pi} \pi\, r^3 \sin\theta\, d\theta$$

$$= \frac{2\pi}{3}\int_0^{\pi} a^3\,(1 + \cos\theta)^3 \sin\theta\, d\theta \qquad \text{from (1)}$$

$$= -\frac{2}{3}\pi\, a^3\int_0^{\pi} c\int_0^{\pi} (1 + \cos\theta)^3\,(-\sin\theta)\, d\theta$$

$$= -\frac{2}{3}\pi\, a^3\left[\frac{(1+\cos\theta)^4}{4}\right]_0^{\pi}, \text{ using power formula}$$

$$= -\frac{1}{6}\pi a^3\,(0 - 24) = \frac{8}{3}\pi a^3.$$

Aliter : *(By Double Integration)* Take a small element r δθ δr at any point P (r, θ) lying within the area of the upper half of the cardioid. Draw PM perpendicular to OX. Then PM = r sin θ. The volume of the elementary ring formed by revolving the element ring formed by revolving the element r δθ δr about OX

$$= 2\pi\,(r\sin\theta)\, r\,\delta\theta\,\delta r$$

$$= 2\pi\, r^2 \sin\theta\,\delta\theta\,\delta r.$$

∴ the required volume formed by revolving the whole cardioid about the initial line

$$= \int_{\theta=0}^{\pi}\int_{r=0}^{a(1+\cos\theta)} 2\pi r^2 \sin\theta\, d\theta\, dr$$

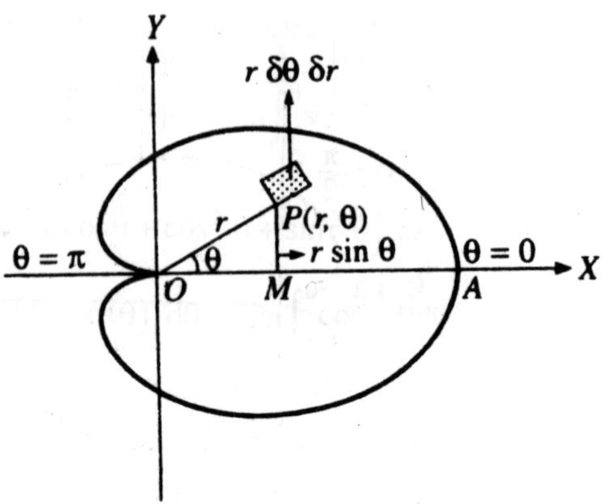

Fig. 3.28

$$= \int_0^{\pi} 2\pi \left[\frac{r^3}{3}\right]_0^{a(1+\cos\theta)} \sin\theta \, d\theta$$

$$= \frac{2\pi}{3} \int_0^{\pi} a^3 (1+\cos\theta)^3 \sin\theta \, d\theta$$

$$= -\frac{2\pi a^3}{3} \int_0^{\pi} (1+\cos\theta)^3 (-\sin\theta) d\theta$$

$$= -\frac{2\pi a^3}{3} \left[\frac{(1+\cos\theta)^4}{4}\right]_0^{\pi}$$

$$= -\frac{2\pi a^3}{3} \cdot \frac{1}{4} [0 - 2^4]$$

$$= \frac{2}{3} \cdot \pi a^3 \cdot \frac{1}{4} \cdot 16$$

$$= \frac{8}{3} \pi a^3.$$

Example 40:

Find the volume of the solid generated by the revolution of the cardioid $r = a(1 + \cos\theta)$ about the initial line.

Solution:

Do your self.

Required volume $= \frac{8}{3} \pi a^3$.

Example 41:

The arc of the cardioid $r = a\,(1 + \cos\theta)$, specified by $-\pi/2 \le \theta \le \pi/2$, is rotated about the line $\theta = 0$, prove that the volume generated is $\frac{5}{2}\pi a^3$.

Solution:

Here the portion B'AB of the cardioid is rotated about the initial line (*i.e.*, x-axis). Obviously the volume generated is the same as is the volume generated by the revolution of the portion AB about x-axis. For the portion AB, θ varies from 0 to $\frac{1}{2}\pi$.

$$\therefore \quad \text{the required volume} = \int_0^{\pi/2} \frac{2}{3}\pi r^3 \sin\theta d\theta$$

$$= \frac{2\pi a^3}{3}\int_0^{\pi/2} (1+\cos\theta)^3 \sin\theta\, d\theta$$

$$= -\frac{1}{6}\pi a^3[1-16] = \frac{15}{6}\pi a^3 = \frac{5}{2}\pi a^3.$$

Example 42:

Show that the volume of the solid formed by the revolution of the curve $r = a + b\cos\theta$ $(a > b)$ about the initial line is $\frac{4}{3}\pi a(a^2 + b^2)$.

Solution:

The given equation of the curve is $r = a + b\cos\theta$ $(a > b)$. ...(1)

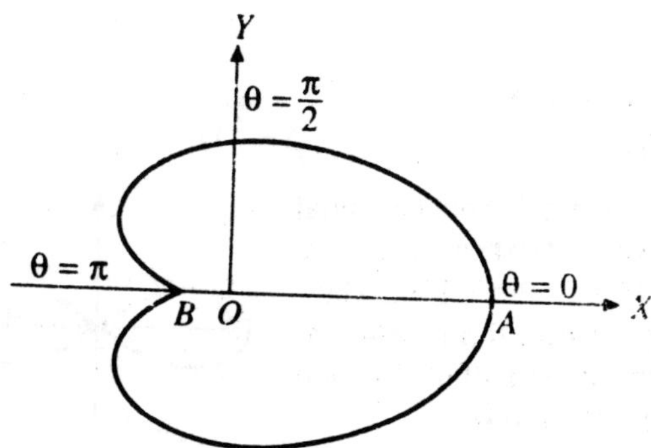

Fig. 3.29

It is symmetrical about the initial line and for the upper half of the curve θ varies from 0 to π.

$\therefore$ the required volume formed by revolving the whole curve about the initial line

$$= \int_0^{\pi} \frac{2}{3}\pi r^2 \sin\theta\, d\theta = \frac{2\pi}{3}\int_0^{\pi} (a + b\cos\theta)^3 \sin\theta d\theta,$$

$$= -\frac{2\pi}{3b}\int_0^{\pi} (a + b\cos\theta)^3 (-b\sin\theta) d\theta$$

$$= -\frac{2\pi}{3b}\left[\frac{(a + b\cos\theta)^4}{4}\right]_0^{\pi} = -\frac{2\pi}{3b}\left[\frac{(a-b)^4}{4} - \frac{(a+b)^4}{4}\right]$$

$$= \frac{\pi}{6b}[(a+b)^4 - (a-b)^4]$$

$$= \frac{\pi}{6b}[(a+b)^2 + (a-b)^2][(a+b)^2 - (a-b)^2]$$

$$= \frac{\pi}{6b} 2(a^2 + b^2).4ab = \frac{4\pi a}{3}(a^2 + b^2).$$

Note: If b = a, then the given curve becomes r = a (1 + cos θ) *i.e.*, a cardioid and hence the volume of the solid generated by the revolution of the cardioid.

r = a (1 + cos θ) about the initial line $= \frac{4}{3}\pi a(a^2 + a^2) = \frac{8}{3}\pi a^3.$

Example 43:

Find the volume of the solid generated by revolving one loop of the lemniscate $r^2 = a^2\cos^2\theta$ *about the line* $\theta = \frac{1}{2}\pi$.

Solution:

The given curve is

$r^2 = a^2\cos^2\theta.$ ...(1)

It is symmetrical about the initial line. We have r = 0 when cos 2θ = 0 *i.e.*, 2θ = ± 1/2π or θ = ± 1/4π. Thus, for one loop θ varies from – π/4to π/4. And for the upper half of one loop θ varies from 0 to 1/4π.

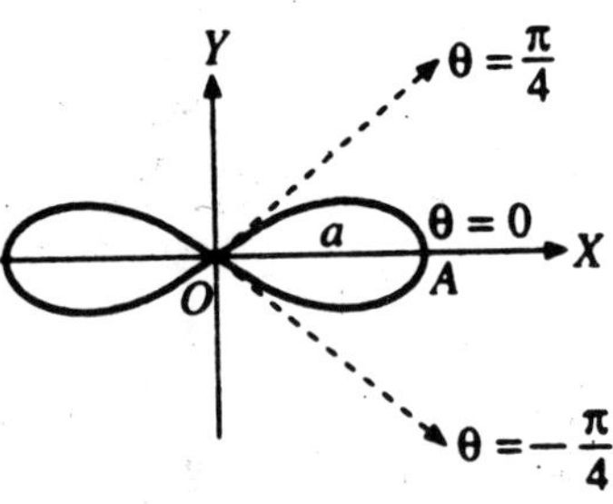

Fig. 3.30

$$= 2\int_0^{\pi/4} \frac{2}{3}\pi r^3 \cos\theta\, d\theta$$

$$= \frac{4\pi}{3}\int_0^{\pi/4} a^3 (\cos 2\theta)^{3/2} \cos\theta\, d\theta,$$

$$= \frac{4\pi a^3}{3}\int_0^{\pi/4}(1-2\sin^2\theta)^{3/2}\cos\theta\, d\theta$$

Now put $\sqrt{2}\sin\theta = \sin\phi$

so that $\sqrt{2}\cos\theta\, d\theta = \cos\phi\, d\phi$.

Also when $\theta = 0$, $\phi = 0$

and when $\theta = \pi/4$, $\phi = \pi/2$.

Then the required volume

$$= \frac{4\pi a^3}{3}\int_0^{\pi/2}(1-\sin^2\phi)^{3/2}\cdot\frac{1}{\sqrt{2}}\cos\phi\, d\phi$$

$$= \frac{4\pi a^3}{3\sqrt{2}}\int_0^{\pi/2}\cos^4\phi\, d\phi$$

$$= \frac{4\pi a^3}{3\sqrt{2}}\cdot\frac{3.1}{4.2}\cdot\frac{\pi}{2} = \frac{\pi^2 a^3}{4\sqrt{2}}.$$

Example 44:

Find the volume of the solid formed by revolving one loop of the curve $r^2 = a^2\cos 2\theta$ about the initial line.

Solution:

The upper half of the loop θ varies from 0 to $\pi/4$. Here the curve is revolving about he initial line (*i.e.*, x-axis).

$$\therefore \text{ the required volume } = \frac{2}{3}\pi\int_0^{\pi/4} r^3\sin\theta\, d\theta$$

$$= \frac{2\pi}{3}\int_0^{\pi/4}\{a\sqrt{(\cos 2\theta)}\}^3\sin\theta\, d\theta \quad [\because r^2 = a^2\cos^2\theta]$$

$$= \frac{2\pi a^3}{3}\int_0^{\pi/4}(2\cos^2\theta - 1)^{3/2}\sin\theta\, d\theta.$$

Put $\sqrt{2}\cos\theta = \sec\phi$

so that $-\sqrt{2}\sin\theta\, d\theta = \sec\phi\tan\phi\, d\phi$.

When $\theta = 0$, $\phi = \pi/4$

and when $\theta = \pi/4$, $\phi = 0$.

$\therefore$ the required volume

$$= \frac{2\pi a^3}{3}\int_{\pi/4}^{0}(\sec^2\phi - 1)^{3/2}\frac{(-\sec\phi\tan\phi)}{\sqrt{2}}d\phi$$

$$= \frac{\sqrt{2}\pi a^3}{3}\int_0^{\pi/4}\tan^4\phi\sec\phi\, d\phi$$

$$= \frac{\sqrt{2}\pi a^3}{3}\int_0^{\pi/4}(\sec^2\phi - 1)^2\sec\phi\, d\phi$$

$$= \frac{\sqrt{2}\pi a^3}{3}\int_0^{\pi/4}(\sec^5\phi - 2\sec^3\phi + \sec\phi)d\phi. \quad ...(1)$$

Also we know the reduction formula

$$\int \sec^n\phi\, d\phi = \frac{\sec^{n-2}\phi\tan\phi}{n-1} + \frac{n-2}{n-1}\int \sec^{n-2}\phi\, d\phi.$$

$$\therefore \quad \int_0^{\pi/4}\sec^5\phi\, d\phi = \left[\frac{\sec^3\phi\tan\phi}{4}\right]_0^{\pi/4} + \frac{3}{4}\int_0^{\pi/4}\sec^3\phi\, d\phi$$

$$= \frac{\sqrt{2}}{2} + \frac{3}{4}\left\{\left[\frac{\sec\phi\tan\phi}{2}\right]_0^{\pi/4} + \frac{1}{2}\int_0^{\pi/4}\sec\phi\, d\phi\right\}$$

$$= \frac{\sqrt{2}}{2} + \frac{3}{4}\left\{\frac{\sqrt{2}}{2} + \frac{1}{2}[\log(\sec\phi + \tan\phi)]_0^{\pi/4}\right\}$$

$$= \frac{\sqrt{2}}{2} + \frac{3\sqrt{2}}{8} + \frac{3}{8}\log(\sqrt{2}+1)$$

$$= \frac{7\sqrt{2}}{8} + \frac{3}{8}\log(\sqrt{2}+1),$$

$$\int_0^{\pi/4}\sec^3\phi\, d\phi = \left[\frac{\sec\phi\tan\phi}{2}\right]_0^{\pi/4} + \frac{1}{2}\int_0^{\pi/4}\sec\phi\, d\phi$$

$$= \frac{\sqrt{2}}{2} + \frac{1}{2}\log(\sqrt{2}+1)$$

and $\int_0^{\pi/4}\sec\phi\, d\phi = \log(\sqrt{2}+1).$

Hence the required volume from (1) is

$$= \frac{\sqrt{2}\pi a^3}{3}\left[\frac{7\sqrt{2}}{8} + \frac{3}{8}\log(\sqrt{2}+1) - 2\left\{\frac{\sqrt{2}}{2} + \frac{1}{2}\log(\sqrt{2}+1)\right\} + \log(\sqrt{2}+1)\right]$$

$$= \frac{\sqrt{2}\pi a^3}{3}\left[\frac{3}{8}\log(\sqrt{2}+1) - \frac{\sqrt{2}}{8}\right]$$

$$= \frac{\pi a^3\sqrt{2}}{24}[3\log(\sqrt{2}+1) - \sqrt{2}].$$

Aliter : The equation of the given curve is

$r^2 = a^2 \cos^2 \theta$ or $r^4 = a^2r^2 (\cos^2 \theta - \sin^2 \theta)$.

Changing to cartesians, the equation becomes $(x^2 + y^2)^2 = a^2 (x^2 - y^2)$ or $y^4 + y^2 (2x^2 + a^2) + x^4 - a^2x^2 = 0$.

Solving for y^2, we have $y^2 = [-(2x^2 + a^2) \pm \sqrt{\{(2x^2 + a^2)^2 - 4 (x^4 - a^2x^2)\}}]/2$.

Neglecting the negative sign because y^2 cannot be –ive, we have

$$y^2 = \frac{-(2x^2 + a^2) + \sqrt{(8a^2x^2 + a^4)}}{2}$$

$$= \frac{-(2x^2 + a^2) + 2\sqrt{2}a\sqrt{\left(x^2 + \frac{1}{8}a^2\right)}}{2}.$$

Now for one loop of the given curve x varies from 0 to a.

$\therefore$ the required volume $= \pi\int_0^a y^2 dx$

$$= \frac{\pi}{2}\int_0^a\left[-2x^2 - a^2 + 2\sqrt{2}a\sqrt{\left(x^2 + \frac{1}{8}a^2\right)}\right]dx$$

$$= \frac{\pi}{2}\int_0^a\left[-\frac{2}{3}x^3 - a^2x + 2\sqrt{2}a\cdot\frac{x}{2}\sqrt{\left(x^2 + \frac{1}{8}a^2\right)}\right.$$

$$\left. + 2\sqrt{2}a\cdot\frac{1}{16}a^2\log\left\{x + \sqrt{\left(x^2 \frac{1}{8}a^2\right)}\right\}\right]_0^a$$

$$= \frac{\pi}{2}\left[-\frac{2}{3}a^3 - a^3 + 2\sqrt{2}a\cdot\frac{a}{2}\cdot\frac{3a}{2\sqrt{2}} + \frac{1}{8}\sqrt{2}a^3\left\{\log\left(a + \frac{3a}{2\sqrt{2}}\right) - \log\frac{a}{2\sqrt{2}}\right\}\right]$$

$$= \frac{\pi}{2}\left[-\frac{5}{3}a^3 + \frac{3}{2}a^3 + \frac{1}{8}\sqrt{2}a^3\log\left\{\frac{a(2\sqrt{2}+3)}{2\sqrt{2}}\cdot\frac{2\sqrt{2}}{a}\right\}\right]$$

$$= \frac{\pi}{2}\left[-\frac{1}{6}a^3 + \frac{1}{8}\sqrt{2}a^3\log(2\sqrt{2}+3)\right]$$

$$= \frac{\pi}{2}\left[-\frac{1}{6}a^3 + \frac{1}{8}\sqrt{2}a^3\log(\sqrt{2}+1)^2\right]$$

$$= \frac{\pi a^3}{2}\left[2\cdot\frac{1}{8}\sqrt{2}\log(\sqrt{2}+1) - \frac{1}{6}\right]$$

$$= \frac{\pi a^3}{2}\left[\frac{1}{4}\sqrt{2}\log(\sqrt{2}+1)-\frac{1}{6}\right]$$

$$= \frac{\pi a^3}{24}[3\sqrt{2}\log(\sqrt{2}+1)-2]$$

$$= \frac{\pi a^3\sqrt{2}}{24}[3\log(\sqrt{2}+1)-\sqrt{2}].$$

Example 45:

The area of the inner loop of the curve r = 1 + 2 cos θ is rotated through two right angles about the inteal. Line. Show that the volume of the solid so formed is π/12.

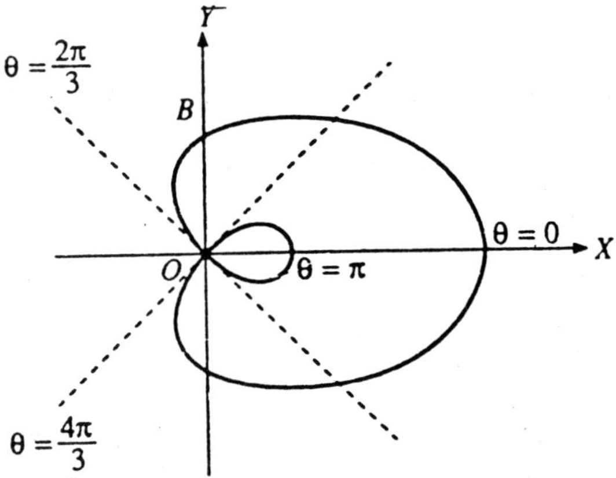

Fig. 3.31

Solution:

The equation of the curve is $r = 1 + 2\cos\theta$. ...(1)

It is symmetrical about the initial line.

Some values of θ and r as gives below:

$\theta = 0$	$\pi/3$	$\pi/2$	$\frac{2\pi}{3}$	π	$\frac{4\pi}{3}$
$r = 3$	3	1	0	-1	0

So the curve consists of an inner loop lying between $\theta = 2\pi/3$ and $\theta = 4\pi/3$ and for the upper half of this inner loop θ varies from θ to $4\pi/3$.

Now the volume generated by the revolution of the whole inner loop through two right angles about the initial line is the same as the volume generated by the revolution through four right angles of the upper half of the inner loop.

$\therefore$ the required volume $= \int_{\pi/3}^{4\pi/3} \frac{2}{3}\pi r^3 \sin\theta \, d\theta$

$= \frac{2\pi}{3}\int_{\pi/3}^{4\pi/3} (1+2\cos\theta)^3 \sin\theta \, d\theta$, substituting for r from (1)

$$= \frac{2\pi}{3}\left[\frac{(1+2\cos\theta)^4}{-8}\right]_{\pi/3}^{4\pi/3} = -\frac{\pi}{12}[0-1] = \frac{\pi}{12}.$$

Example 46:

Show that if the area lying within the cardioid $r = 2a(1 + \cos\theta)$ and without the parabola $r(1 + \cos\theta) = 2a$ revolves about the initial line, the volume generated is $18\pi a^3$.

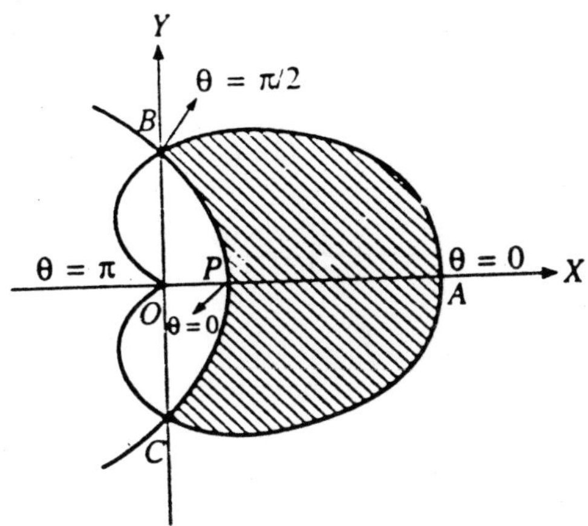

Fig. 3.32

Solution:

The equation of the cardioid is $r = 2a(1 + \cos\theta)$, ...(1)

and that of the parabola is $r = 2a/(1 + \cos\theta)$. ...(2)

Equating the values of r from (1) and (2), we get

$$2a(1 + \cos\theta) = 2a/(1 + \cos\theta)$$

or $(1 + \cos\theta)^2 = 1$

or $\cos\theta(\cos\theta + 2) = 0.$

Now $\cos\theta \neq -2$.

Therefore $\cos\theta = 0$

i.e., $\theta = \pi/2, -\pi/2$.

Thus, the curves (1) and (2) intersect where $\theta = \pi/2$ and $\pi = -\pi/2$.

Also both the curve are symmetrical about the initial line (*i.e.*, x-axis). The required volume is generated by revolving the upper half of the shaded area about the initial line.

$\therefore$ the required volume = Volume generated by the revolution of the area OABO of the cardioid – volume generated by the revolution of the area OPBO of the parabola

$$= \frac{2\pi}{3}\int_0^{\pi/2} r^3 \sin\theta\,\delta\theta - \frac{2\pi}{3}\int_0^{\pi/2} r^3 \sin\theta\, d\theta$$

(for cardioid) (for parabola)

$$= \frac{2\pi}{3}\int_0^{\pi/2}\left[8a^3(1+\cos\theta)^3 - \frac{8a^3}{(1+\cos\theta)^3}\right]\sin\theta\, d\theta$$

$$= \frac{-16\pi a^3}{3}\int_0^{\pi/2}[(1+\cos\theta)^3 - (1+\cos\theta)^{-3}](-\sin\theta)d\theta$$

$$= \frac{-16\pi a^3}{3}\left[\frac{(1+\cos\theta)^4}{4} - \frac{(1+\cos\theta)^{-2}}{-2}\right]_0^{\pi/2} \quad \text{using power formula}$$

$$= \frac{-16\pi a^3}{3}\left[\frac{1}{4}(1-16) + \frac{1}{2}\left(1-\frac{1}{4}\right)\right]$$

$$= \frac{-16}{3}\pi a^3\left[-\frac{15}{4} + \frac{3}{8}\right]$$

Example 47:

Find the curved surface of a hemisphere of radius a.

Solution:

A hemisphere is generated by the revolution of a quadrant of a circle about of its bounding radii.

Let the equation of the circle be $x^2 + y^2 = a^2$. ...(1)

Let the hemisphere be formed by revolving about x-axis the arc of the circle (1) lying in the first quadrant.

Differentiating (1), w.r.t. x, we get $2x + 2y\,(dy/dx) = 0$

or $dy/dx = -x/y$.

Therefore $$\frac{ds}{dx} = \sqrt{\left\{1+\left(\frac{dy}{dx}\right)^2\right\}} = \sqrt{\left\{1+\frac{x^2}{y^2}\right\}}$$

$$= \sqrt{\left\{\frac{y^2+x^2}{y^2}\right\}} = \sqrt{\left(\frac{a^2}{y^2}\right)}, \text{ from (1)}$$

For the arc of the circle (1) lying in the first quadrant x varies from 0 to a.

∴ the required surface

$$= 2\pi\int_{x=0}^{a} y\,ds = 2\pi\int_0^a y\frac{ds}{dx}\cdot dx$$

$$= 2\pi\int_0^a y\cdot\frac{a}{y}dx = 2\pi\int_0^a a\,dx = 2\pi\, a[x]_0^a$$

$$= 2\pi a.a = 2\pi a^2.$$

Example 48:

Find the surface of a sphere of radius a.

Solution:

Suppose the sphere is generated by the revolution of a semi-circle of radius a about its bounding diameter (say x-axis).

Let the equation of the circle be $x^2 + y^2 = a^2$, ...(1)

the center being the origin.

Then ds/dx = a/y.

Also for the semi-circle, x varies from – a to a.

∴ the required surface

$$= 2\pi\int_{x=-a}^{a} y\,ds = 2\pi\int_{-a}^{a} y\frac{ds}{dx}dx$$

$$= 2\pi\int_{-a}^{a} y\cdot\frac{a}{y}dx$$

$$= 2\pi\int_{-a}^{a} a\,dx = 2\pi a[x]_{-a}^{a}$$

$$= 2\pi a\,(a + a) = 4\pi a^2.$$

Example 49:

Show that the surface of the spherical zone contained between two parallel planes is 2πah where a is the radius of the sphere and h the distance between the places.

Solution:

Let the sphere be generated by the revolution about the x-axis of the circle $x^2 + y^2 = a^2$.

Let the two parallels places bounding the spherical zone be formed by the revolution of the lines x = b and x = b + h.

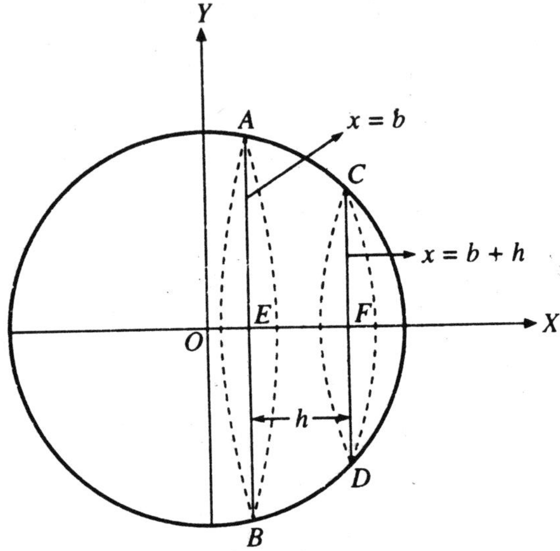

Fig. 3.33

Then the required surface is generated by the revolution of the arc AC about x-axis.

We get

$$\frac{ds}{dx} = \frac{a}{y}.$$

$\therefore$ the required surface

$$= \int_b^{b+h} 2\pi y \frac{ds}{dx} dx$$

$$= 2\pi \int_b^{b+h} y \cdot \frac{a}{y} dx$$

$$= 2\pi a \int_b^{b+h} dx = 2\pi a [x]_b^{b+h}$$

$$= 2\pi a\ (b + h - h) = 2pab.$$

Example 50:

Find the area of the surface formed by the revolution of the parabola $y^2 = 4ax$ about the x-axis by the arc from the vertex to one end of the laws rectum.

Solution:

The given parabola is $y^2 = 4ax$. ...(1)

Differentiating (1) w.r.t. x, we get $2y\,(dy/dx) = 4a$ or $dy/dx = 2a/y$.

$$\therefore \qquad \frac{ds}{dx} = \sqrt{\left\{1+\left(\frac{dy}{dx}\right)^2\right\}} = \sqrt{\left(1+\frac{4a^2}{y^2}\right)}$$

$$= \sqrt{\left(\frac{y^2+4a^2}{y^2}\right)}$$

$$= \frac{\sqrt{(4ax+4a^2)}}{y}, \qquad \text{from (1)}$$

$$= \frac{2\sqrt{a}\sqrt{(x+a)}}{y}. \qquad ...(2)$$

For the given arc from the vertex (0, 0) to one end of the latus recuts, say the end (a, 2a), x varies from 0 to a.

$\therefore$ the required surface

$$= \int_{x=0}^{a} 2\pi y \frac{ds}{dx}\,dx$$

$$= 2\pi \int_0^a y \cdot \frac{2\sqrt{a}\sqrt{(x+a)}}{y}\,dx, \text{ from (2)}$$

$$= 4\pi\sqrt{a}\int_0^a (x+a)^{1/2}\,dx = 4\pi\sqrt{a}\cdot\left[\frac{2}{3}(x+a)^{3/2}\right]_0^a$$

$$= \frac{8\pi\sqrt{a}}{3}[(2a)^{3/2} - a^{3/2}] = \frac{8\pi a^2}{3}[2\sqrt{2}-1].$$

Example 51:

Find the curved surface of the solid generated by the revolution, about the x-axis of the area bounded by the x-axis the parabola $y^2 = 4ax$ and the ordinate $x = h$.

Solution:

As above example. The required curved surface

$$= \int_{x=0}^{h} 2\pi y \frac{ds}{dx} dx$$

$$= 2\pi \int_0^h y \cdot \frac{2\sqrt{a}\sqrt{(x+a)}}{y} dx$$

$$= 4\pi\sqrt{a} \int_0^h (x+a)^{1/2} dx$$

$$= \frac{8\pi}{3}\sqrt{a}[(h+a)^{3/2} - a^{3/2}].$$

Example 52:

Find the surface generated by the revolution of on arc of the catenary $y = c \cosh (x/c)$ about the axis of x.

Solution:

The given curve is, $y = c \cos h\ (x/c)$. ...(1)

Differentiating (1) w.r.t. x, we get

$$\frac{dy}{dx} = c \sin h \frac{x}{c} \cdot \frac{1}{c}$$

$$= \sin h \frac{x}{c}.$$

$$\therefore \quad \frac{ds}{dx} = \sqrt{\left\{1 + \left(\frac{dy}{dx}\right)^2\right\}}$$

$$= \sqrt{\left\{1 + \sinh^2 \frac{x}{c}\right\}} = \cosh \frac{x}{c} \quad \text{...(2)}$$

If the arc be measured from the vertex (x = 0) to any point (x, y), then the required surface formed by the revolution of this arc about x-axis

$$= \int_{x=0}^{x} 2\pi y \frac{ds}{dx} dx$$

$$= 2\pi \int_0^x c \cosh \frac{x}{c} \cdot \cosh \frac{x}{c} dx, \text{ from (1) and (2)}$$

$$= \pi c \int_0^x 2\cosh^2 \frac{x}{c} dx$$

$$= \pi c \int_0^x \left[1 + \cosh \frac{2x}{c}\right] dx$$

$$= \pi c \left[x + \frac{c}{2} \sinh \frac{2x}{c}\right]_0^x$$

$$= \pi c\left[x + \frac{c}{2}\sin\frac{2x}{c}\right]$$

$$= \pi c\left[x + c\sinh\frac{x}{c}\cosh\frac{x}{c}\right].$$

Example 53:

Find the surface generated by the revolution of the curve y = c cosh (a/c) and about the x-axis, between the places x = a and x = b.

Solution:

Do your self. The required surface

$$= 2\pi\int_{x=a}^{b} y\frac{ds}{dx}dx$$

$$= \pi c\left[x + \frac{c}{2}\sinh\left(\frac{2x}{c}\right)\right]_a^b$$

$$= \pi c\left[(b-a) + \frac{c}{2}\sinh\frac{2b}{c} - \frac{c}{2}\sinh\frac{2a}{c}\right].$$

Example 54:

For a catenary y = a cosh (x/a), prove that aS = 2V = πa (ax + sy), where s is the length of the arc from the vertex, S and V are respectively the area of the curved surface and volume of the solid generated by the revolution of the arc about x-axis.

Solution:

The given equation of catenary is y = a cosh (x/a). ...(1)

The vertex of the catenary (1) is the point (0, a).

Differentiating (1) w.r.t. x we get dy/dx = sinh (x/a).

$$\frac{ds}{dx} = \sqrt{\left\{1 + \left(\frac{dy}{dx}\right)^2\right\}}$$

$$= \sqrt{\left\{1 + \sinh^2\frac{x}{a}\right\}} = \cosh\frac{x}{a}$$

or $$ds = \cosh(x/a)\,dx.$$

If s is the length of the arc from the vertex (x = 0) to any point (x, y) on the catenary, then we have

$$s = \int_0^x \cosh\left(\frac{x}{a}\right)dx$$

$$= \left[a \sinh \frac{x}{a} \right]_0^a = a \sinh \frac{x}{a}. \qquad ...(2)$$

Now S = area of the curved surface of the solid generated by the revolution of the arc about x-axis

$$= \int_0^x 2\pi y \frac{ds}{dx} dx = \pi a \left[x + \frac{a}{2} \sinh \frac{2x}{a} \right] \qquad ...(3)$$

Also V = the volume generated by the revolution of the arc about x-axis

$$= \int_0^x \pi y^2 dx = \pi \int_0^x a^2 \cosh^2 \frac{x}{a} dx$$

$$= \frac{\pi a^2}{2} \int_0^x 2 \cosh^2 \frac{x}{a} dx$$

$$= \frac{\pi a^2}{2} \int_0^x \left[1 + \cosh \frac{2x}{a} \right] dx$$

$$= \frac{\pi a^2}{2} \left[x + \frac{a}{2} \sinh \frac{2x}{a} \right]_0^x$$

$$= \frac{\pi a^2}{2} \left[x + \frac{a}{2} \sinh \frac{2x}{a} \right]. \qquad ...(4)$$

From (3), $$aS = \pi a^2 \left[x + \frac{1}{2} a \sinh(2x/a) \right]. \qquad ...(5)$$

From (4), $$2V = \pi a^2 \left[x + \frac{1}{2} a \sinh(2x/a) \right] = aS. \qquad ...(6)$$

Form (5) and (6), we have aS = 2V.

Also from (2), $\pi a\,(ax + sy) = \pi a\,[ax + a^2 \sinh(x/a) \cosh(x/a)]$,

$[\because\ y = a \cosh(x/a)]$

$$= \pi a \left[ax + \frac{a^2}{2} \cdot 2 \sinh \frac{x}{a} \cosh \frac{x}{a} \right]$$

$$= \pi a^2\, [x + (a/2) \sinh(2x/a)[= aS, \qquad \text{from (5).}$$

Hence $aS = 2V = \pi a\,(ax + sy)$.

Example 55:

Prove that the surface of the prolate spheroid formed by the revolution of the ellipse of eccentricity e about its major axis is equal to 2 × area of the ellipse × $[\sqrt{(1 - e^2)} + (1/e) \sin^{-1} e]$.

Solution:

Let the equation of the ellipse be $\dfrac{x^2}{a^2}+\dfrac{y^2}{b^2}=1,$ ...(1)

The x-axis being the major axis so that a > b.

The parametric equations of (1) are x = a cosh t, y = b sin t.

$\therefore$ dx/dt = – a sin t and dy/dt = b cos t.

We have $$\frac{ds}{dt}=\sqrt{\left\{\left(\frac{dx}{dt}\right)^2+\left(\frac{dy}{dt}\right)^2\right\}}$$

$$=\sqrt{(a^2\sin^2 t+b^2\cos^2 t)}$$

$=\sqrt{\{a^2\sin^2 t+a^2(1-e^2)\cos^2 t.\}}$, [$\because$ for the ellipse $b^2=a^2(1-e^2)$]

$=a\sqrt{(1-e^2\cos^2 t)}$. ...(2)

Now the ellipse (1) is symmetrical about y-axis and for the arc of the ellipse lying in the first quadrant t varies from 0 to π/2. At the point (a, 0) we have t = 0 and at the point (0, b) we have t = π/2.

Hence the required surface S formed by the revolution of the ellipse (1) about the x-axis

$$=2\int 2\pi y\,ds \text{ between the suitable limits}$$

$$=4\pi\int_0^{\pi/2} y\frac{ds}{dt}dt$$

$$=4\pi\int_0^{\pi/2} b\sin t.a\sqrt{(1-e^2\cos^2 t)}dt,$$

[$\because$ y = b sin t and ds/dt = $a\sqrt{(1-e^2\cos^2 t)}$, from (2)]

$$=4\pi ab\int_0^{\pi/2}\sin t\sqrt{(1-e^2\cos^2 t)}dt.$$

Put e cos t = z
so that – e sin t dt = dz.
When t = 0, z = e

and when $t=\dfrac{1}{2}\pi$, z = 0.

$$\therefore \quad S=-4\pi ab\int_e^0\frac{1}{e}\sqrt{(1-z^2)}dz$$

$$=\frac{4\pi ab}{e}\int_0^e\sqrt{(1-z^2)}dz$$

$$=\frac{4\pi ab}{e}\left[\frac{z}{2}\sqrt{(1-z^2)}+\frac{1}{2}\sin^{-1}z\right]_0^e$$

$$= \frac{4\pi ab}{e}\left[\frac{e}{2}\sqrt{(1-e^2)} + \frac{1}{2}\sin^{-1} e\right]$$

$$= 2\pi ab\ [\sqrt{(1 - e^2)} + (1/e) \sin^{-1} e]$$

$$= 2 \times \text{area of the ellipse} \times [\sqrt{(1-e^2)}+(1/e) \sin^{-1} e].$$

Remark:

The solid of revolution formed by revolving an ellipse about its minor axis is called an *oblate spheroid.*

Example 56:

Find the surface of the solid generated by the revolution of the ellipse $x^2 + 4y^2 = 16$ *about its major axis.*

Solution:

The given ellipse is $x^2 + 4y^2 = 16$. ...(1)

The equation (1) of the ellipse can be written as

$$\frac{x^2}{16} + \frac{y^2}{4} = 1.$$

Comparing it with the standard form of the equation of the ellipse $x^2/a^2 + y^2/b^2 = 1$, we see that $a = 4$, $b = 2$ and the major axis is along the x-axis. So we have to revolve the curve (1) about x-axis.

Differentiating (1) w.r.t. x, we get

$$2x + 8y\frac{dy}{dx} = 0$$

or $$\frac{dy}{dx} = -\frac{x}{4y}.$$

$$\therefore \quad \frac{ds}{dx} = \sqrt{\left\{1+\left(\frac{dy}{dx}\right)^2\right\}}$$

$$= \sqrt{\left\{1+\frac{x^2}{16y^2}\right\}}$$

$$= \sqrt{\left(\frac{16y^2+x^2}{16y^2}\right)}$$

$$= \sqrt{\{(64 - 4x^2) + x^2\}}/4y = \sqrt{(64 - 3x^2)}/4y. \quad ...(2)$$

Now the ellipse (1) is symmetrical about both axes and for the arc of the ellipse lying in the first quadrant x varies from 0 to 4.

∴ the required surface = 2 × surface generated by the revolution of the arc in the first quadrant

$$= 2\int_{x=0}^{4} 2\pi y \frac{ds}{dx} dx$$

$$= 4\pi \int_0^4 y \cdot \frac{\sqrt{(64 - 3x^2)}}{4y} dx, \qquad \text{from (2)}$$

$$= \pi \int_0^4 \sqrt{(64 - 3x^2)} dx$$

$$= \pi \sqrt{3} \int_0^4 \sqrt{\left\{\left(\frac{8}{\sqrt{3}}\right)^2 - x^2\right\}} dx$$

$$= \pi \sqrt{3} \left[\frac{x}{2}\sqrt{\left\{\left(\frac{8}{\sqrt{3}}\right)^2 - x^2\right\}} + \frac{1}{2} \cdot \frac{64}{3} \sin^{-1}\left(\frac{x\sqrt{3}}{8}\right)\right]_0^4$$

$$= \pi \sqrt{3} \left[2\sqrt{\left(\frac{64}{3} - 16\right)} + \frac{32}{3} \sin^{-1}\left(\frac{\sqrt{3}}{2}\right)\right]$$

$$= \pi \sqrt{3} \left[\frac{8}{\sqrt{3}} + \frac{32}{3} \cdot \frac{\pi}{3}\right], \quad \left[\because \sin^{-1}\left(\frac{\sqrt{3}}{2}\right) = \frac{\pi}{3}\right]$$

$$= 8\pi \left[1 + \frac{4\pi}{3\sqrt{3}}\right].$$

Example 57:

Find the surface of the solid formed by the revolution, about the axis of y, of the part of the curve $ay^2 = x^3$ from $x = 0$ to $x = 4a$ which is above the x-axis.

Solution:

The given curve is $ay^2 = x^3$. ...(1)

Differentiating (1) w.r.t. x, we get

$$2ay \frac{dy}{dx} = 3x^2 \text{ or } \frac{dy}{dx} = \frac{dx^2}{2ay} s$$

$$\therefore \qquad \frac{ds}{dx} = \sqrt{\left\{1 + \left(\frac{dy}{dx}\right)^2\right\}} = \sqrt{\left\{1 + \frac{9x^4}{4a^2y^2}\right\}}$$

$$= \sqrt{\left\{1 + \frac{9x^4}{4a.x^3}\right\}}, \text{ from (1)}$$

$$= \sqrt{\left(1 + \frac{9x}{4a}\right)} = \frac{1}{2\sqrt{a}}\sqrt{(4a + 9x)}.$$

$\therefore$ the required surface

$$= \int_{x=0}^{4a} 2\pi x \, ds = \int_0^{4a} 2\pi x \frac{ds}{dx} dx$$

$$= \int_0^{4a} 2\pi x \cdot \frac{1}{2\sqrt{a}} \sqrt{(4a + 9x)} dx$$

$$= \frac{\pi}{\sqrt{a}} \int_0^a x \sqrt{(4a + 9x)} dx.$$

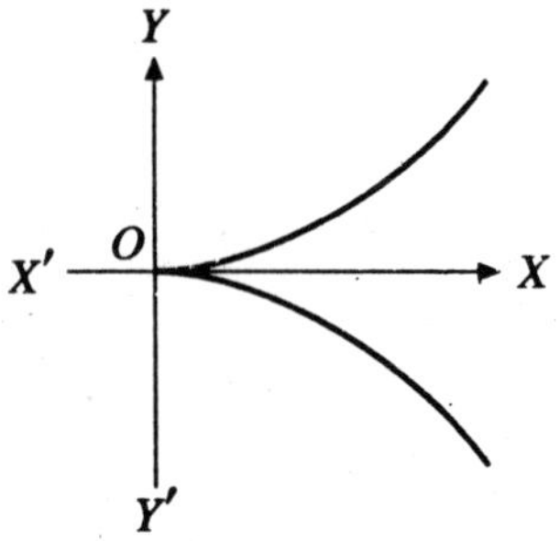

Fig. 3.34

Put $4a + 9x = t^2$

so that $9 \, dx = 2t \, dt.$

When $x = 0, t = \sqrt{(4a)}$

and when $x = 4a, t = \sqrt{(40a)}.$

$\therefore$ the required surface $= \dfrac{\pi}{\sqrt{a}} \displaystyle\int_{\sqrt{(4a)}}^{\sqrt{(40a)}} \frac{t^2 - 4a}{9} \cdot t \cdot \frac{2t \, dt}{9}$

$$= \frac{2\pi}{81\sqrt{a}} \int_{\sqrt{(4a)}}^{\sqrt{(40a)}} (t^4 - 4at^2) dt$$

$$= \frac{2\pi}{81\sqrt{a}} \left[\frac{1}{5}t^5 - \frac{4}{3}at^3\right]_{\sqrt{(4a)}}^{\sqrt{(40a)}}$$

$$= \frac{128}{1215} \pi a^2 [125\sqrt{(10)} + 1].$$

Example 58:

The curve $ay^2 = x^3$ revolves about the axis of y; find the surface area and the volume generated between the planes perpendicular to the axis of revolution at the origin and through the point where $27y = 8a$.

Solution:

The given curve is $ay^2 = x^3$. ...(1)

$$\frac{ds}{dx} = \sqrt{\left(1 + \frac{9x}{4a}\right)}.$$

Let us find the value of x at the point on the curve (1) where $27y = 8a$.

Putting y = 8a/27 in (1), we get $64a^3/27^2 = x^3$ or x = 4a/9.

$\therefore$ the lines perpendicular to the axis of revolution, at the origin and through the point where 27y = 8a meet the curve (1) at the points where x = 0 and x = 4a/9.

$\therefore$ the required surface $= \int_{x=0}^{4a/9} 2\pi x \frac{ds}{dx} dx$

$$= 2\pi \int_0^{4a/9} x \cdot \sqrt{\left(\frac{4a+9x}{4a}\right)} dx = \frac{\pi}{\sqrt{a}} \int_0^{4a/9} x\sqrt{(4a+9x)}dx.$$

Now put $4a + 9x = t^2$

so that 9 dx = 2t dt.

Also when x = 0, t = 2 $\sqrt{a}$

and when x = 4a/9, t = 2 $\sqrt{2}$. $\sqrt{a}$.

$\therefore$ the required surface

$$= \frac{\pi}{\sqrt{a}} \int_{2\sqrt{a}}^{2\sqrt{2}\sqrt{a}} \frac{(t^2 - 4a)}{9} \cdot t \cdot \frac{2t\,dt}{9}$$

$$= \frac{2\pi}{81\sqrt{a}} \int_{2\sqrt{a}}^{2\sqrt{2}\sqrt{a}} (t^4 - 4at^2)dt = \frac{2\pi}{81\sqrt{a}} \left[\frac{t^5}{5} - \frac{4at^3}{3}\right]_{2\sqrt{a}}^{2\sqrt{2}\sqrt{a}}$$

$$= \frac{2\pi}{81\sqrt{a}} \left[\left\{\frac{(2\sqrt{2}\sqrt{a})^5}{5} - \frac{4a}{3}(2\sqrt{2}\sqrt{a})^3\right\} - \left\{\frac{(2\sqrt{a})^5}{5} - \frac{4a(2\sqrt{a})^3}{3}\right\}\right]$$

$$= \frac{2\pi}{81\sqrt{a}} \left[\frac{128\sqrt{2}.a^2\sqrt{a}}{5} - \frac{64\sqrt{2}a^2\sqrt{a}}{3} - \frac{32a^2\sqrt{a}}{5} + \frac{32a^2\sqrt{a}}{3}\right]$$

$$= \frac{2\pi a^2}{81 \times 15}[384\sqrt{2} - 320\sqrt{2} - 96 + 160]$$

$$= \frac{2\pi a^2}{81 \times 15}[64\sqrt{2} + 64]$$

$$= \frac{2\pi a^2}{1215}[64\sqrt{2} + 64].$$

Example 59:

Find the area of the surface of revolution formed by revolving the curve r = 2a cos θ about the initial line.

Solution:

The given curve is r = 2a cos θ, ...(1)

which is clearly a circle of radius a passing through the pole and having diameter through the pole as initial line. At O, r = 0 and so (1) gives $\theta = \pi/2$ at O.

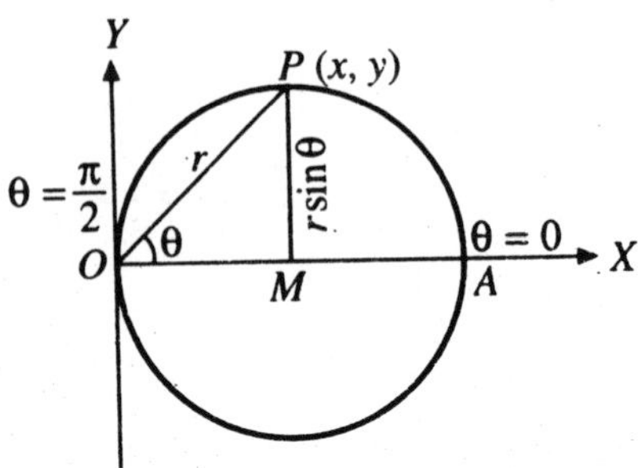

Fig. 3.35

Differentiating (1) w.r.t. θ, we have $dr/d\theta = -2a \sin\theta$.

$$\therefore \quad \frac{ds}{d\theta} = \sqrt{\left\{r^2 + \left(\frac{dr}{d\theta}\right)^2\right\}}$$

$$= \sqrt{\{4a^2 \cos^2\theta + 4a^2 \sin^2\theta\}}$$

$$= 2a\sqrt{(\cos^2\theta + \sin^2\theta)} = 2a.$$

The given curve is revolved about the initial line (*i.e.*, the x-axis) and for the upper half of the curve, θ varies from 0 to π/2.

$$\therefore \quad \text{the required surface} = \int_0^{\pi/2} 2\pi y \frac{ds}{d\theta} d\theta, \text{ where } y = r \sin\theta$$

$$= 2\pi \int_0^{\pi/2} r \sin\theta . 2a d\theta = 4a\pi \int_0^{\pi/2} 2a \cos\theta \sin\theta \, d\theta.$$

$$= 8\pi a^2 \int_0^{\pi/2} \sin\theta \cos\theta \, d\theta$$

$$= 8\pi a^2 \left[\frac{\sin^2\theta}{2}\right]_0^{\pi/2} = 8\pi a2 \left(\frac{1}{2} - 0\right) = 4\pi a2.$$

Example 60:

Find the surface of the solid generated by the revolution of the lemniscate $r^2 = a^2 \cos^2\theta$ about the initial line.

Solution:

The given curve is $r^2 = a^2 \cos 2\theta$. ...(1)

Differentiating (1) w.r.t. θ, we get

$$2r\frac{dr}{d\theta} = -\ 2a^2 \sin 2\theta$$

or
$$\frac{dr}{d\theta} = \frac{-a^2 \sin 2\theta}{r}.$$

∴
$$\frac{ds}{d\theta} = \sqrt{\left\{r^2 + \left(\frac{dr}{d\theta}\right)^2\right\}}$$

$$= \sqrt{\left\{a^2 \cos 2\theta + \frac{a^4 \sin^2 2\theta}{r^2}\right\}}$$

$$= \frac{1}{r}\sqrt{\{r^2 . a^2 \cos 2\theta + a^4 \sin^2 2\theta\}}$$

$$= \frac{1}{r}\sqrt{\{a^4 \cos^2 2\theta + a^4 \sin^2 2\theta\}}, \qquad [\because r^2 = a^2 \cos 2\theta]$$

$$= a^2/r. \qquad ...(2)$$

The given curve is symmetrical about the initial line and about the pole.

Putting r = 0 in (1), we get cos 2θ = 0

giving $2\theta = \pm\frac{1}{2}\pi$

i.e., $\theta = \pm\ \frac{1}{4}\pi$

Therefore one loop of the curve lies between $\theta = -\frac{1}{4}\pi$ and $\theta = \frac{1}{4}\pi$.

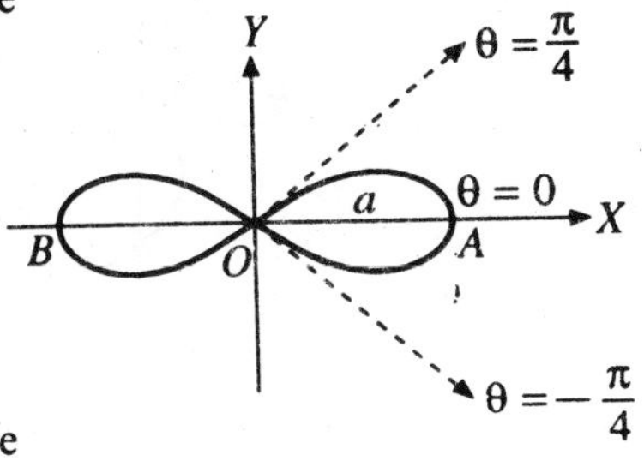

Fig. 3.36

There are two loops in the curve and for the upper half of one of these two loops θ varies from 0 to $\frac{1}{4}\pi$.

∴ the required surface = 2 × the surface generated by the revolution of one loop

$$= 2.\int_0^{\pi/4} 2\pi y \frac{ds}{d\theta} d\theta, \text{ where } y = r \sin\theta$$

$$= 4\pi\int_0^{\pi/4} r \sin\theta \cdot \frac{a^2}{r} d\theta, \qquad \text{from (2)}$$

$$= 4\pi a^2 \int_0^{\pi/4} \sin\theta \, d\theta$$

$$= 4\pi a^2 [-\cos\theta]_0^{\pi/4}$$

$$= 4\pi a^2 [-(1/\sqrt{2}) + 1]$$

$$= 4\pi a^2 [1 - (1/\sqrt{2})].$$

Example 61:

Find the surface of the said formed by the revolution of the cardioid $r = a (1 + \cos\theta)$ about the initial line.

Solution:

The given curve is $r = a (1 + \cos\theta)$. ...(1)

It is symmetrical about the initial line and for the upper half of the curve, θ varies from 0 to π.

Differentiating (1) w.r.t. θ, we get $dr/d\theta = a(-\sin\theta) = -a\sin\theta$.

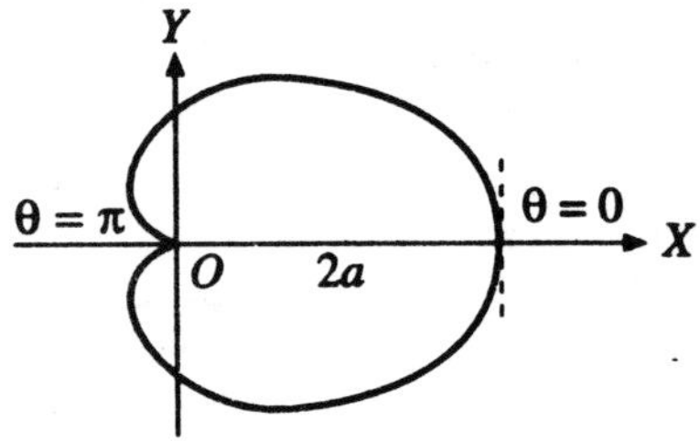

Fig. 3.37

$$\therefore \quad \frac{ds}{d\theta} = \sqrt{\left[r^2 + \left(\frac{dr}{d\theta}\right)^2\right]}$$

$$= \sqrt{[a^2 (1 + \cos\theta)^2 + a^2 \sin^2\theta]}$$

$$= a\sqrt{[2(1 + \cos\theta)]} = 2a\cos\frac{1}{2}\theta. \qquad ...(2)$$

$\therefore$ the required surface $= \int_0^{\pi} 2\pi y \frac{ds}{d\theta} d\theta$, where $y = r\sin\theta$

$$= \int_0^{\pi} 2\pi . r\sin\theta . 2a\cos\frac{1}{2} d\theta$$

$$= 2\pi \int_0^{\pi} a(1 + \cos\theta)\sin\theta\cos\frac{1}{2}\theta \, d\theta \text{ from (1)}$$

$$= 4\pi a^2 \int_0^{\pi} 2\cos^2\frac{\theta}{2}\cdot 2\sin\frac{\theta}{2}\cos\frac{\theta}{2}\cdot\cos\frac{\theta}{2}\,d\theta$$

$$= 16\pi a^2 \int_0^{\pi/2} \cos^4\frac{\theta}{2}\sin\frac{\theta}{2}\,d\theta$$

$$= 16\pi a^2 \int_0^{\pi/2} (\cos^4 t \sin t).2\,dt, \text{ putting } \frac{\theta}{2}$$

$= t$ so that $d\theta = 2dt$

$$= 32\pi a^2 \int_0^{\pi/2} \cos^4 t \sin t\,dt$$

$$= 32\pi a^2 \cdot \frac{3.1.1}{5.3.1} = \frac{32\pi a^2}{5}.$$

Example 62:

Find the surface of the solid generated by the revolution of the curve $r = a\,(1 - \cos\theta)$ about the initial line.

Solution:

Do your self.

The required surface is $(3/5)\,\pi a^2$.

Example 63:

A circular arc revolves about its chord. Find the area of the surface generated, when 2α is the angle subtended by the arc at the centre.

Solution:

Let the parametric equations of the circle be $x = a\cos\theta$, $y\ a\sin\theta$, θ being the parameter. ...(1)

Take any point $P\,(a\cos\theta,\ a\sin\theta)$ on the circular arc ABC which is symmetrical about the x-axis and which subtends an angle 2α at the centre O so that $< AOB = \alpha$.

We have $OD = OA\cos\alpha = a\cos\alpha$.

Draw PM perpendicular from

Then $PM = ON - OD = a\cos\theta - a\cos\alpha$. ...(2)

For the upper half of the arc to be rotated *i.e.*, for the arc BA, θ varies from 0 to α.

Also $$\frac{ds}{d\theta} = \sqrt{\left\{\left(\frac{dx}{d\theta}\right)^2 + \left(\frac{dy}{d\theta}\right)^2\right\}}$$

$$= \sqrt{\{a^2\sin^2\theta + a^2\cos^2\theta\}} = a.$$

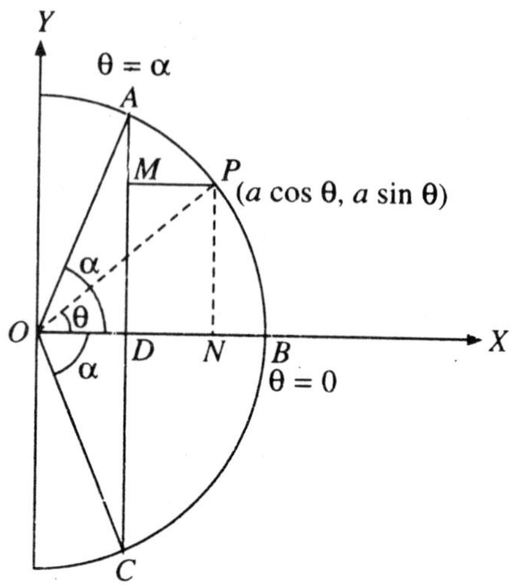

Fig. 3.38

∴ the required surface = 2 × surface generated by the revolution of the arc BA about the chord AC

$$= 2\times\int_0^{\alpha} 2\pi(\text{PM})\frac{ds}{d\theta}\,d\theta$$

$$= 4\pi\int_0^{\alpha}(a\cos\theta - a\cos\alpha).a.d\theta, \qquad \text{from (2)}$$

$$= 4\pi a^2\left[\sin\theta - \theta\cos\alpha\right]_0^{\alpha}$$

$$= 4\pi a^2[\sin\alpha - \alpha\cos\alpha]\cdot$$

Example 64:

A quadrant of a circle of radius a revolves about its chord Show that the surface of the spindle generated is

$$2\pi a^2\sqrt{2}\left(1-\frac{1}{4}\pi\right).$$

Solution:

Let us take the quadrant of the circle in such a way that is placed symmetrically about the x-axis. The quadrant of the circle subtends and angle $\pi/2$ at the centre.

By taking $2\alpha = \pi/2$ and taking $\alpha = \pi/4$. Thus, the required result is obtained by putting $a = \pi/4$.

Example 65:

Find the area of the surface generated if an arch of the cycloid $x = a(\theta - \sin\theta)$, $y = a(1 - \cos\theta)$ revolves about the line $y = 2a$.

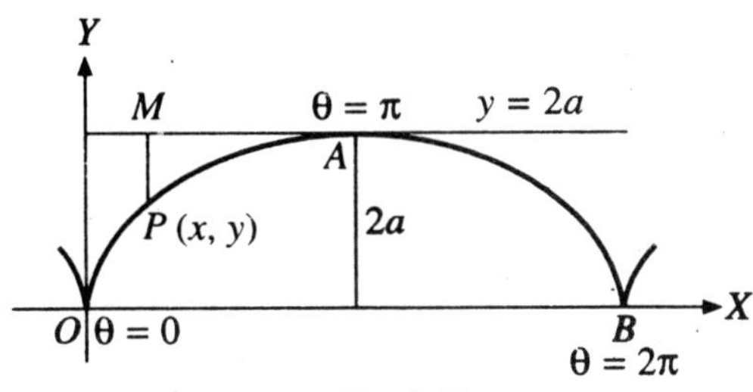

Fig. 3.39

Solution:

The given parametric equation of the cycloid are

$$x = a(\theta - \sin\theta),\ y = a(1 - \cos\theta). \qquad ...(1)$$

Differentiating (1) w.r.t. θ,

we get $dx/d\theta = a(1 - \cos\theta)$ and $dy/d\theta = a\sin\theta$.

$$\frac{ds}{d\theta} = \sqrt{\left\{\left(\frac{dx}{d\theta}\right)^2 + \left(\frac{dy}{d\theta}\right)^2\right\}}$$

$$= \sqrt{\{a^2(1 - \cos\theta)^2 + a^2\sin^2\theta\}}$$

$$= a\sqrt{\{2(1 - \cos\theta)\}} = a\sqrt{\{2.2\sin^2(\theta/2)\}} = 2a\sin(\theta/2).$$

Take P (x, y) as any point on the are OA. Draw PM perpendicular to the line $y = 2a$, which is tangent to the cycloid at the vertex A. Then PM $= 2a - y = 2a - a(1 - \cos\theta) = a(1 + \cos\theta)$.

Also the given cycloid is symmetrical about a line which is perpendicular to x-axis and which meets the curve at the point A where $\theta = \pi$. For the arc OA, θ varies from 0 to π.

∴ the required surface

= 2 × the surface formed by the revolution of the arc OA about the line $y = 2a$

$$= 2 \times \int_0^{\pi} 2\pi(PM)\frac{ds}{d\theta}\,d\theta$$

$$= 4\pi\int_0^{\pi} a(1 + \cos\theta).2a\sin\frac{\theta}{2}\,d\theta$$

$$= 8\pi a^2 \int_0^{\pi} 2\cos^2 \frac{\theta}{2} \cdot \sin \frac{\theta}{2}\, d\theta$$

$$= -\ 32\pi a^2 \int_0^{\pi} \cos^2 \frac{\theta}{2}\left(-\frac{1}{2}\sin \frac{\theta}{2}\right) d\theta,$$

Making adjustment for the application of the power formula

$$= -\ 32\pi a^2 \left[\frac{\cos^3 \frac{1}{2}\theta}{3}\right]$$

$$= -\ \frac{32}{3}\pi a^2 [0-1] = \frac{32}{3}\pi a^2.$$

Example 66:

The lemniscate $r^2 = a^2 \cos 2\theta$ revolves about a tangent at the pole. Show that surface of the solid generated is $4\pi a^2$.

Solution:

The given curve is $r^2 = a^2 \cos 2\theta$. ...(1)

We get $ds/d\theta = a^2/r$.

Putting $r = 0$ in (1),

we get $\cos 2\theta = 0$ giving $2\theta = \pm\ 1/2\pi$

i.e., $\theta = \pm\ 1/4\pi$.

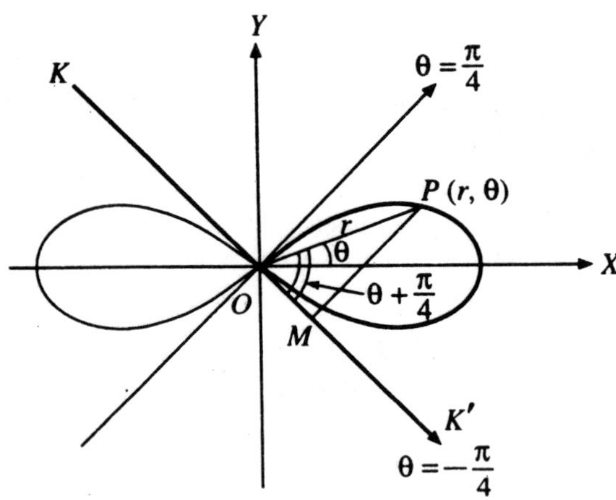

Fig. 3.40

Therefore, on loop of the curve lies between $\theta = -1/4\pi$ and $\theta = 1/4\pi$. The curve consists of two loops and both the lines $\theta = -1/4\pi$ and $\theta = 1/4\pi$ are tangents at the pole. Let the curve be revolved about the line KOK' which is a tangent at the pole.

Take any point P (r, θ) on the curve and draw PM perpendicular to the axis fo rotation KOK'. Then $< POM = \frac{1}{4}\pi + \theta$ and $MP = OP \sin\left(\frac{1}{4}\pi + \theta\right)$

$= r\sin\left(\frac{1}{4}\pi + \theta\right)$.

Als for one loop θ varies from $-\frac{1}{4}\pi$ to $\frac{1}{4}\pi$.

∴ the required surface = 2 × surface generated by one loop

$$= 2 \times \int_{-\pi/4}^{\pi/4} 2\pi(PM)\frac{ds}{d\theta}d\theta$$

$$= 4\pi\int_{-\pi/4}^{\pi/4} r\sin\left(\frac{1}{4}\pi + \theta\right)\cdot\frac{a^2}{r}d\theta \quad \left[\because \frac{ds}{d\theta} = \frac{a^2}{r}, PM = r\sin\left(\frac{1}{4}\pi + \theta\right)\right]$$

$$= 4\pi a^2\int_{-\pi/4}^{\pi/4} \sin\left(\frac{1}{4}\pi + \theta\right)d\theta = 4\pi a^2\left[-\cos\left(\frac{1}{4}\pi + \theta\right)\right]_{-\pi/4}^{/4}$$

$$= 4\pi a^2 [0 + 1] = 4\pi a^2.$$

Example 67:

Find the volume and surface-area of the anchor-ring generated by the revolution of a circle of radius a about an axis in its own plane distant b from its centre (b > a).

Solution:

Here the given curve (circle) does not intersect the axis of rotation, so Pappus theorem can be applied. In this case

A = area of the region of the closed curve

= area of the circle of radius a

$= \pi a^2$

and l = length of the arc of the curve = circumference of the circle

$= 2\pi.$

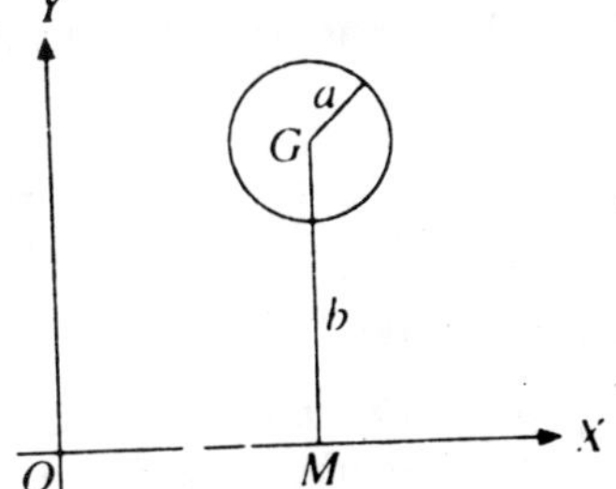

Fig. 3.41

As the centroid of the area of a circle and also of its circumference lies at the centre, so $\overline{y} = b$ in both the cases and hence the length of the path described by the C.G. $= 2\pi b$.

Now by Pappus theorem, the required volume of the anchor-ring

= area of the circle × circumference of the circle generated by the centroid $= \pi a^2 \,.\, 2\pi b = 2\pi^2 a^2 b$.

And the surface area of the anchor-ring

= arc length of the circle × circumference of the circle generated by the centroid $= 2\pi a.\ 2\pi b = 4\pi^2 ab$.

Example 68:

Find the position of the centroid of a semi-circular area.

Solution:

Clearly, the C.G. of the sexi-circular area will lie somewhere on the radius which is perpendicular to the bounding diameter. Let the distance of the centroid from the center O be y. Also a sphere will be generated by the rotation of the semi-circular area about the bounding diameter.

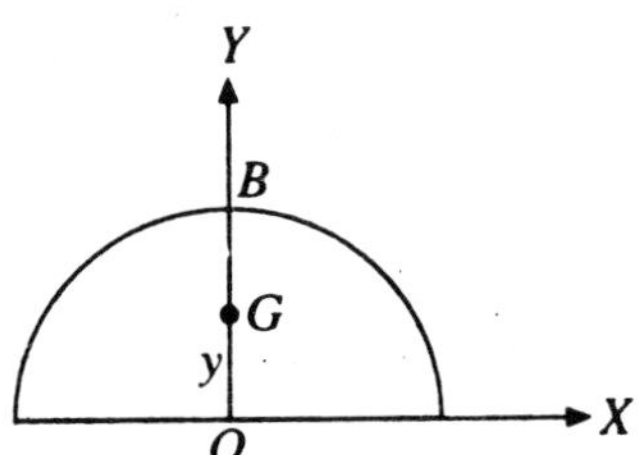

Fig. 3.42

∴ by Pappus theorem, volume of the solid of revolution

= area of semi-circle × Circumference of the circle generated by the centroid of this area.

Hence $\frac{4}{3}\pi a^3 = \frac{1}{2}\pi a^2,\ 2\pi y$,

where a is the radius of the semi-circle

or $$y = \frac{1}{2\pi}\cdot\frac{\frac{4}{3}\pi a^3}{\frac{1}{2}\pi a^2} = \frac{4a}{3\pi}$$

Example 69:

Show that the volume generated by the revolution of the ellipse $x^2/a^2 + y^2/b^2 = 1$ about the line $x = 2a$ is $4\pi^2 a^2 b$.

Solution:

Area of the given ellipse is πab. ...(1)

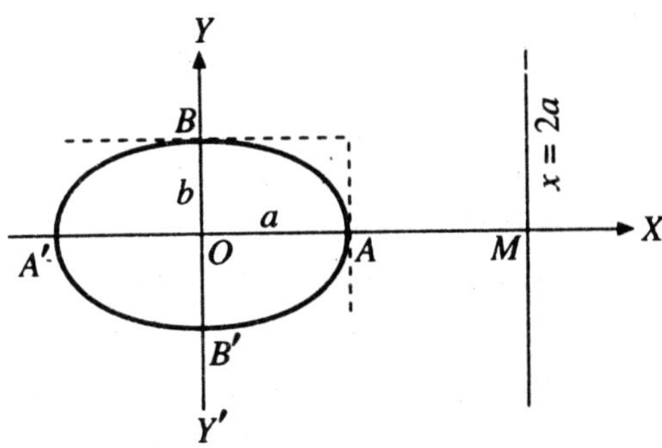

Fig. 3.43

The C.G. Of the ellipse will describe a circle of radius 2a when revolved about the line x = 2a. Hence the length of the arc described by the

C.G. = 2π (2a) = $4\pi a$.

$\therefore$ by Pappus theorem the required volume

= area of the ellipse $\times$ length of the arc described by its C.G.

= $\pi ab.\ 4\pi a = 4\pi^2 a^2 b$.

Example 70:

Show that the volume generated by the revolution of an ellipse having semi-axes a and b about a tangent at the vertex is $2\pi^2 a^2 b$ or $2\pi^2 ab^2$.

Solution:

Area of the given ellipse = πab. ...(1)

The C.G. of the ellipse will describe a circle of radius a when revolved about the tangent at A or a circle of radius b when revolved about the tangent at B.

Hence, the length of the arc described by the C.G. Will be 2π. 2a *i.e.*, $4\pi a$ or 2π. 2b *i.e.*, $4\pi b$.

$\therefore$ by Pappus theorem, the required volume

(i) when revolved about the tangent at the vertex A

= $\pi ab.\ 2\pi a = 2\pi^2 a^2 b$,

(ii) when removed about the tangent at the vertex B

= $\pi ab.\ 2\pi b = 2\pi^2 ab^2$.

Example 71:

The part of the parabola $y^2 = 4ax$ cut off by the latus rectum revolves about the tangent at the vertex. Find the curved surface of the reel thus generated.

Solution:

The given parabola is $y^2 = 4ax$. ...(1)

Differentiating (1) w.r.t. x, we get $dy/dx = 2a/y$.

$$\therefore \quad \frac{ds}{dx} = \sqrt{\left\{1+\left(\frac{dy}{dx}\right)^2\right\}} = \sqrt{\left\{1+\frac{4a^2}{y^2}\right\}}$$

$$= \sqrt{\left\{1+\frac{4a^2}{4ax}\right\}} = \sqrt{\left(\frac{x+a}{x}\right)}.$$

The required curved surface is generated by the revolution of the arc LOL' (LSL' is the latus rectum), about the tangent at the vertex *i.e.*, y-axis. The curve is symmetrical about x-axis and for the arc OL, x varies from 0 to a.

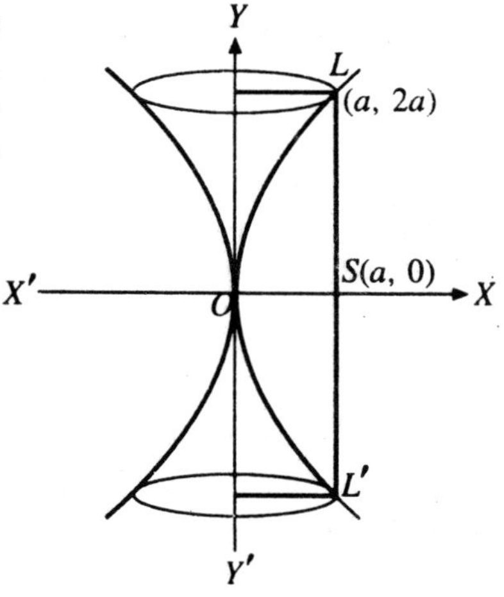

Fig. 3.44

$\therefore$ the required surface

$$= 2\int_{x=0}^{a} 2\pi x \frac{ds}{dx} dx$$

$$= 4\pi \int_0^a x\sqrt{\left(\frac{x+a}{x}\right)} dx$$

$$= 4\pi \int_0^a \sqrt{(x^2+ax)}dx$$

$$= 4\pi \int_0^a \sqrt{\left\{\left(x+\frac{a}{2}\right)^2-\left(\frac{a}{2}\right)^2\right\}}dx$$

$$= 4\pi \left[\frac{1}{2}\left(x+\frac{a}{2}\right)\sqrt{(x^2+ax)} - \frac{1}{2}\cdot\frac{a^2}{4}\log\left\{\left(x+\frac{a}{2}\right)+\sqrt{(x^2+ax)}\right\}\right]_0^a$$

$$\left[\because \int \sqrt{(x^2-a^2)}dx = \frac{1}{2}x\sqrt{(x^2-a^2)} - \frac{1}{2}a^2\log\{x+\sqrt{(x^2-a^2)}\}\right]$$

$$= 4\pi \left[\frac{1}{2}\cdot\frac{3}{2}a\,a\sqrt{2} - \frac{1}{8}a^2\log\left\{\frac{3}{2}a + a\sqrt{2}\right\} + \frac{1}{8}a^2\log\left(\frac{1}{2}a\right)\right]$$

$$= 4\pi\left[\frac{3}{4}a^2\sqrt{2} - \frac{1}{8}a^2\log\left\{\left(\frac{3}{2}a + a\sqrt{2}\right)\Big/\left(\frac{1}{2}a\right)\right\}\right]$$

$$= 4\pi\left[3\sqrt{2} - \frac{1}{2}\log(3 + 2\sqrt{2})\right]$$

$$= \pi a^2\left[3\sqrt{2} - \frac{1}{2}\log(\sqrt{2} + 1)^2\right]$$

$$= \pi a^2\,[3\sqrt{2} - \log(\sqrt{2} + 1)].$$

Example 72:

Find the surface of the solid generated by revolution of the astroid $x^{2/3} + y^{2/3} = a^{2/3}$ *or* $x = a\cos^3 t$, $y = a\sin^3 t$ *about the x-axis.*

Solution:

The parametric equations of the curve are $x = a\cos^3 t$, $y = a\sin^3 t$.

$$\therefore \quad \frac{dx}{dt} = -3a\cos^3 t\sin t$$

and $$\frac{dy}{dt} = 3a\sin^2 t\cos t.$$

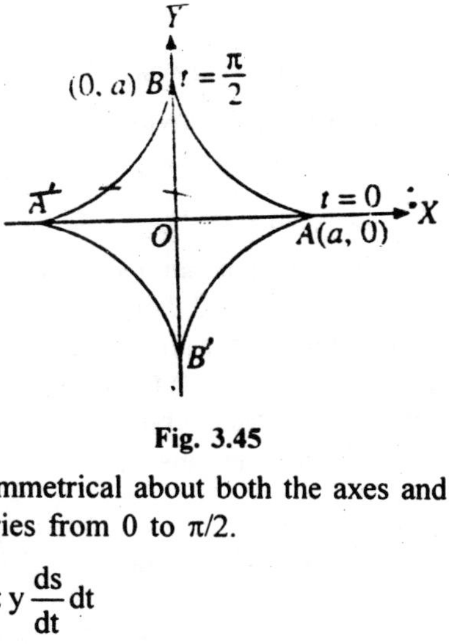

Hence $$\frac{ds}{dt} = \sqrt{\left[\left(\frac{dx}{dt}\right)^2 + \left(\frac{dy}{dt}\right)^2\right]}$$

$$= \sqrt{[9a^2\cos^4 t\sin^2 t + 9a^2\sin^4 t\cos^2 t]}$$

$$= \sqrt{[9a^2\sin^2 t\cos^2 t(\cos^2 t + \sin^2 t)]}$$

$$= 3a\sin t\cos t.$$

Fig. 3.45

Also the given curve (astroid) is symmetrical about both the axes and for the curve in the first quadrant, t varies from 0 to $\pi/2$.

$$\therefore \quad \text{the required surface} = 2\int_{t=0}^{\pi/2} 2\pi y\frac{ds}{dt}dt$$

$$= 4\pi\int_0^{\pi/2} a\sin^3 t.3a\sin t\cos t\,dt$$

$$= 12\pi a\int_0^{\pi/2}\sin^4\cos t\,dt$$

$$= 12\pi a^2\left[\frac{\sin^5 t}{5}\right]_0^{\pi/2} = 12\pi a^2\left[\frac{1}{5} - 0\right] = \frac{12\pi a^2}{5}.$$

Example 73:

Find the surface area of the solid generated by revolving the cycoid $x = a(\theta - \sin\theta)$, $y = a(1 - \cos\theta)$ about the x-axis.

Solution:

The given parametric equations of the cycloid are

$$x = a(\theta - \sin\theta),\ y = a(1 - \cos\theta). \qquad ...(1)$$

$$\therefore \qquad \frac{dx}{d\theta} = a(1 - \cos\theta) \text{ and } \frac{dy}{d\theta} = a\sin\theta.$$

Hence
$$\frac{ds}{d\theta} = \sqrt{\left[\left(\frac{dx}{d\theta}\right)^2 + \left(\frac{dy}{d\theta}\right)^2\right]}$$
$$= \sqrt{[a^2(1 - \cos\theta)^2 + a^2\sin^2\theta]}$$
$$= a\sqrt{[1 - 2\cos\theta + \cos^2\theta + \sin^2\theta]}$$
$$= a\sqrt{[2(1 - \cos\theta)]}$$
$$= a\sqrt{[2.2\sin^2(\theta/2)]} = 2a\sin(\theta/2). \qquad (2)$$

We have $y = 0$ when $1 - \cos\theta = 0$

i.e., $\cos\theta = 1$ giving $\theta = 0$ and 2π.

When $\theta = 0$, $x = 0$

and when $\theta = 2\pi$, $x = 2a\pi$.

Also y is maximum when $\cos\theta = -1$ *i.e.*, $\theta = \pi$ and then $y = 2a$ and $x = a\pi$. Thus for one arch of the given curve θ varies from 0 to 2π and this arch is symmetrical about the line $x = a\pi$ which meets the curve at the point $\theta = \pi$.

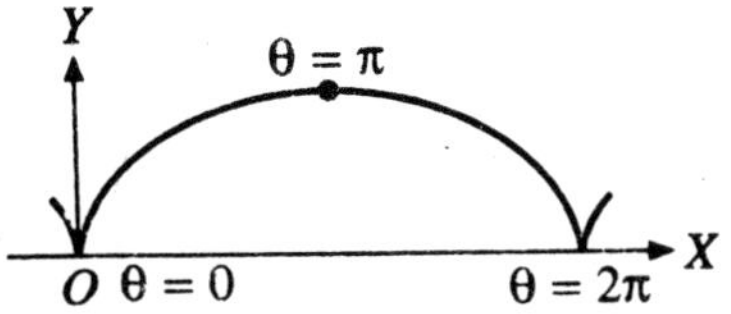

Fig. 3.46

$$\therefore \quad \text{the required surface} = 2\int_0^{\pi} 2\pi y \frac{ds}{d\theta} d\theta$$

$$= 4\pi\int_0^{\pi} a(1 - \cos\theta).\ 2a\sin(\theta/2)\,d\theta, \text{ from (1) and (2)}$$

$$= 8\pi a^2\int_0^{\pi} 2\sin^2\frac{\theta}{2} d\theta \sin\frac{\theta}{2} = 16\pi a^2\int_0^{\pi}\sin^3\frac{\theta}{2} d\theta$$

$$= 16\pi a^2\int_0^{\pi/2}\sin^3 t.2dt,$$

putting $\frac{\theta}{2} = t$

so that $d\theta = 2\,dt$

$$= 32\pi a^2 \cdot \frac{2}{3.1} = \frac{64\pi a^2}{3}.$$

Example [illegible].

Find the area of the surface generated by revolving an arch of the cycloid x = a (θ + sin θ), y = a (1 − cos θ) about the tangent at the vertex.

Solution:

The given parametric equations of the cycloid are

$$x = a\,(\theta + \sin\theta),\ y = a\,(1 - \cos\theta). \qquad ...(1)$$

$$\therefore \qquad \frac{dx}{d\theta} = a\,(1 + \cos\theta) \text{ and } \frac{dy}{d\theta} = a \sin\theta.$$

Hence
$$\frac{ds}{d\theta} = \sqrt{\left[\left(\frac{dx}{d\theta}\right)^2 + \left(\frac{dy}{d\theta}\right)^2\right]}$$
$$= \sqrt{[a^2\,(1 + \cos\theta)^2 + a^2 \sin^2\theta]}$$
$$= a\,\sqrt{[1 + 2\cos\theta + \cos^2\theta + \sin^2\theta]}$$
$$= a.\,\sqrt{[2\,(1 + \cos\theta)]}$$
$$= a.\,\sqrt{\{2.2\cos^2(\theta/2)\}} = 2a\cos(\theta/2). \qquad ...(2)$$

Also for one arch of the given curve, θ varies form −π to π and this arch is symmetrical about the y-axis which meets the curve at the point θ = 0.

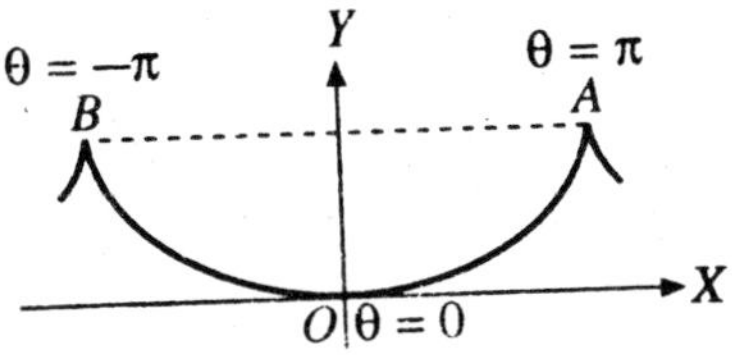

Fig. 3.47

∴ the required surface

$$= 2\int_0^{\pi} 2\pi y \frac{ds}{d\theta}\,d\theta$$

$$= 4\pi\int_0^{\pi} a(1 - \cos\theta).2a\cos(\theta/2)d\theta, \text{ from (1) and (2)}$$

$$= 8\pi a^2\int_0^{\pi} 2\sin^2\frac{\theta}{2}\cdot\cos\frac{\theta}{2}\,d\theta$$

$$= 16\pi a^2\int_0^{\pi}\sin^2\frac{\theta}{2}\cos\frac{\theta}{2}\,d\theta$$

$$= 16\pi a^2\int_0^{\pi/2}\sin^2 t\cos t.2dt, \text{ putting } \frac{\theta}{2} = t \text{ so that } d\theta = 2dt,$$

$$= 32\pi a^2\int_0^{\pi/2}\sin^2 t\cos t\,dt = 32\pi a^2\cdot\frac{1}{3.1} = \frac{32\pi a^2}{3}.$$

Example 75:

The portion between the consecutive cusps of the cycloid $x = a(\theta + \sin\theta)$, $y = a(1 + \cos\theta)$ is revolved about the x-axis. Prove that the area of the surface so formed is to the area of the cycloid as 64:9.

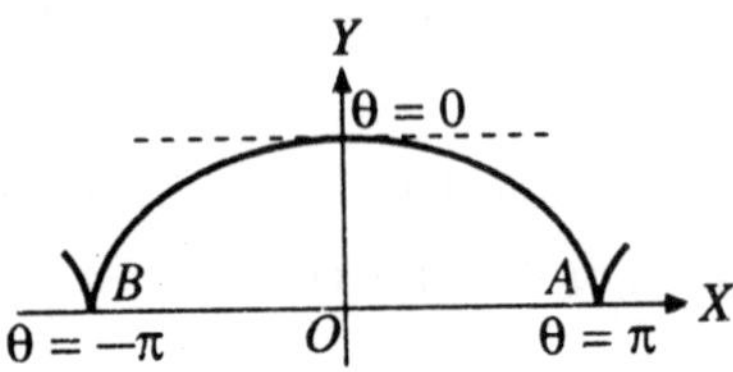

Fig. 3.48

Solution:

The given parametric equations of the cycloid are

$$x = a(\theta + \sin\theta),\ y = a(1 + \cos\theta), \qquad ...(1)$$

$$\therefore \quad \frac{dx}{d\theta} = a(1 + \cos\theta),\ \frac{dx}{d\theta} = a\sin\theta.$$

Hence

$$\frac{ds}{d\theta} = \sqrt{\left\{\left(\frac{dx}{d\theta}\right)^2 + \left(\frac{dy}{d\theta}\right)^2\right\}}$$

$$= \sqrt{[a^2(1 + \cos\theta)^2 + a^2\sin^2\theta]}$$

$$= a\sqrt{[1 + 2\cos\theta + \cos^2\theta + \sin^2\theta]}$$

$$= a\sqrt{[2(1 + \cos\theta)]}$$

$$= a\sqrt{[2.2\cos^2(\theta/2)]} = 2a\cos(\theta/2). \qquad ...(2)$$

For one arch of the given curve (*i.e.*, for the portion between two successive cusps) θ varies from $-\pi$ to π. Also this arch is symmetrical about the y-axis which meets the arch at the point where $\theta = 0$. The base of the given cycloid is the axis of x.

$\therefore$ the surface S generated by the revolution of the cycloid about the x-axis $= 2\int_0^\pi 2\pi y \frac{ds}{d\theta} d\theta$

$$= 4\pi\int_0^\pi a(1 + \cos\theta).2a\cos(\theta/2)d\theta, \text{ from (1) and (2)}$$

$$= 8\pi a^2\int_0^\pi 2\cos^2\frac{\theta}{2}\cos\frac{\theta}{2}d\theta$$

$$= 16\pi a^2 \int_0^{\pi} \cos^3 \frac{\theta}{2} d\theta$$

$$= 16\pi a^2 \int_0^{\pi/2} (\cos^3 t).2dt, \text{ putting } \theta/2 = t \text{ so that } d\theta = 2dt$$

$$= 32\pi a^2 \cdot \frac{2}{3.1} = \frac{64\pi a^2}{3}.$$

Also the area A of the given cycloid

$$= 2\int_0^{\pi} y \frac{dx}{d\theta} d\theta = 2\int_0^{\pi} a(1+\cos\theta).a(1+\cos\theta)d\theta$$

$$= 2a^2 \int_0^{\pi} (1+\cos\theta)^2 d\theta$$

$$= 2a^2 \int_0^{\pi} \left(2\cos^2 \frac{\theta}{2}\right)^2 d\theta$$

$$= 8a^2 \int_0^{\pi} \cos^4 \frac{\theta}{2} d\theta$$

$$= 8a^2 \int_0^{\pi/2} (\cos^4 t).2dt, \text{ putting } t = \theta/2$$

$$= 16a^2 \cdot \frac{3.1}{4.2} \cdot \frac{\pi}{2} = 3\pi a^2. \quad ...(4)$$

From (3) and (4), we get

$$\text{The required ratio} = \frac{S}{A} = \frac{\frac{64}{3}\pi a^2}{3\pi a^2} = \frac{64}{9}.$$

Example 76:

Prove that the surface area of the solid generated by the revolution, about the x-axis of the of the curve $x = t^2$, $y = t - \frac{1}{3}t^3$ *is* 3π.

Solution:

The given eqautions of the curve are $x = t^2$, $y = t - \frac{1}{3}t^3$. ...(1)

$\therefore$ $dx/dt = 2t$ and $dy/dt = 1 - t^2$.

Hence $$\frac{ds}{dt} = \sqrt{\left[\left(\frac{dx}{dt}\right)^2 + \left(\frac{dy}{dt}\right)^2\right]}$$

$$= \sqrt{[(2t)^2 + (1 - t^2)^2]}$$

$$= \sqrt{[4t^2 + 1 - 2t^2 + t^4]}$$

$$= \sqrt{(1 + t^2)^2} = (1 + t^2). \qquad ...(2)$$

Putting y = 0 in (1), we get $t - \frac{1}{3}t^3 = 0$ which gives t = 0 or t = ± √3. For the upper half of the loop y is positive and so for the upper half of the loop t from 0 to √3.

$\therefore$ the required surface $= \int_0^{\sqrt{3}} 2\pi\, y \frac{ds}{dt}\, dt$

$$= 2\pi \int_0^{\sqrt{3}} \left(t - \frac{1}{3}t^3\right)(1 + t^2)\,dt$$

$$= 2\pi \int_0^{\sqrt{3}} \left(t + \frac{2}{3}t^3 - \frac{1}{3}t^5\right)dt$$

$$= 2\pi \left[\frac{t^2}{2} + \frac{2}{3}\cdot\frac{t^4}{4} - \frac{1}{3}\frac{t^6}{6}\right]_0^{\sqrt{3}}$$

$$= 2\pi \left[\frac{t^2}{2} + \frac{t^4}{6} - \frac{t^6}{18}\right]_0^{\sqrt{3}}$$

$$= 2\pi \left[\frac{3}{2} + \frac{9}{6} - \frac{27}{18}\right] = 3\pi.$$

Example 77:

Prove that the surface of the solid generated by the revolution of the tractirx $x = a \cos t + \frac{1}{2} a \log \tan^2 \frac{1}{2} t$, $y = a \sin t$ *about its asymptote is equal to the surface of a sphere of radius a.*

Solution:

The given tractrix is x = a cos t + 1/2 a log $\tan^2$ 1/2 t, y = a sin t.

$$\therefore \qquad \frac{dx}{dt} = -a \sin t + a \frac{\sec^2 \frac{1}{2} t}{\tan \frac{1}{2} t} \cdot \frac{1}{2}$$

$$= a \left[-\sin t + \frac{1}{2 \sin \frac{1}{2} t \cos \frac{1}{2} t}\right]$$

$$= a\left(-\sin t + \frac{1}{\sin t}\right)$$

$$= a\,\frac{(-\sin^2 t + 1)}{\sin t} = \frac{a\cos^2 t}{\sin t}$$

and $\qquad dy/dt = a \cos t.$

Hence $\qquad \dfrac{ds}{dt} = \sqrt{\left\{\left(\dfrac{dx}{dt}\right)^2 + \left(\dfrac{dy}{dt}\right)^2\right\}}$

$$= \sqrt{\left\{\frac{a^2\cos^4 t}{\sin^2 t} + a^2\cos^2 t\right\}} = \frac{a\cos t}{\sin t}.$$

The given curve is symmetrical about both the axes and the asymptote is the line y = 0 *i.e.*, x-axis. For the arc of the curve lying in the second quadrant t varies from 0 to $\frac{1}{2}\pi$.

$\therefore$ the required surface $= 2.\int_0^{\pi/2} 2\pi y \dfrac{ds}{dt} dt$

$$= 4\pi\int_0^{\pi/2} a\sin t.\frac{a\cos t}{\sin t} dt$$

$$= 4\pi a^2\int_0^{\pi/2} \cos t\, dt$$

$$= 4\pi a^2[\sin t]_0^{\pi/2} = 4pa2$$

= the surface of a sphere of radius a.

Example 78:

Prove That the surface of the oblate spheroid formed by the revolution of the ellipse of the semi-major axis a and eccentricity e is

$$2\pi a^2\left[1 + \frac{1-e^2}{2e}\log\left(\frac{1+e}{1-e}\right)\right].$$

Solution:

Let the parametric equations of the ellipse be

$x = a\cos t,\ y = b\sin t$, where $b^2 = a^2(1 - e^2)$. ...(1)

$\therefore$ dx\dt = –a sin t and dy/dt = b cos t.

$\therefore \qquad \dfrac{ds}{dt} = \sqrt{\left\{\left(\dfrac{dx}{dt}\right)^2 + \left(\dfrac{dy}{dt}\right)^2\right\}}$

$$= \sqrt{\{a^2 \sin^2 t + b^2 \cos^2 t\}}$$
$$= \sqrt{\{a^2 \sin^2 t + a^2 (1 - e^2) \cos^2 t\}}$$
$$= a \sqrt{(1 - e^2 \cos^2 t)} \qquad ...(2)$$

The ellipse is symmetrical about both the axes and for the arc of the ellipse lying in the first quadrant t varies from 0 to $\pi/2$.

We have to revolve the ellipse about its minor axis which is the y-axis.

$\therefore$ the required surface $= 2\int_0^{\pi/2} 2\pi x \frac{ds}{dt} dt$

$$= 4\pi \int_0^{\pi/2} a \cos t . a \sqrt{(1 - e^2 \cos^2 t)} dt$$

$$= 4\pi a^2 \int_0^{\pi/2} \sqrt{(1 - e^2 + e^2 \sin^2 t)} \cos t \, dt$$

$$= \frac{4\pi a^2}{e} \int_0^e \sqrt{[(1 - e^2) + z^2]} dz, \text{ putting } e \sin t = z \text{ so that } e \cos t \, dt = dz$$

$$= \frac{4\pi a^2}{e} \int_0^e \left[\frac{z}{2} \sqrt{\{(1 - e^2) + z^2\}} + \frac{1}{2}(1 - e^2) \log\{z + \sqrt{(1 - e^2 + z^2)}\} \right]_0^e$$

$$= \frac{2\pi a^2}{e} [e + (1 - e^2) \log(e + 1) - (1 - e^2) \log \sqrt{(1 - e^2)}]$$

$$= \frac{2\pi a^2}{e} \left[e + (1 - e^2) \log \frac{1 + e}{\sqrt{(1 - e^2)}} \right]$$

$$= 2\pi a^2 \left[1 + \frac{1 - e^2}{e} \log \sqrt{\left(\frac{1 + e}{1 - e} \right)} \right]$$

$$= 2\pi a^2 \left[1 + \frac{1 - e^2}{2e} \log \frac{1 + e}{1 - e} \right].$$

Example 79:

Find the volume of the ring generated by the revolution of an ellipse of eccentricity $1/\sqrt{2}$ about a straight line parallels to the minor axis and situated at distance from the centre equal to three times the major axis.

Solution:

Let a be the semi-major axis of the ellipse. Then its semi-minor axis

$$b = a \sqrt{(1 - e^2)} = a \sqrt{\left(1 - \frac{1}{2}\right)} = a/\sqrt{2}. \qquad [\because e = 1/\sqrt{2}].$$

$\therefore$ area of the ellipse $= \pi a = \pi a.\ (a/\sqrt{2}) = \pi a^2/\sqrt{2}$.

Distance of the C.G. of the ellipse from the axis of revolution is 3.2a 6a, (given).

As the ellipse revolves about the given line its C.G. will describe a circle of radius 6a whose perimeter will be

$$= 2\pi.\ 6a = 12\pi a.$$

Now by Pappus theorem, the required volume

= area of the ellipse × length of the arc described by its C.G.

$$= (\pi a^2/\sqrt{2}).\ 12\pi a = 12\pi^2 a3/\sqrt{2}$$

$$= 6\sqrt{2}\ \pi^2 a^3,$$ where a is the semi-major axis.

Example 80:

The loop of the curve $2ay^2 = x(x - a)^2$ revolves about the straight line $y = a$. Find the volume of the solid generated.

Solution:

The given curve is $2ay^2 = x(x - a)^2$ ...(1)

The curve (1) is symmetrical about the x-axis and the loop lies between $x = 0$ and $x = a$.

Differentiating (1) w.r.t. x, we get $4ay\ (dy/dx) = 2x(x - a) + (x - a)^2 = 3x^2 - 4ax + a^2$.

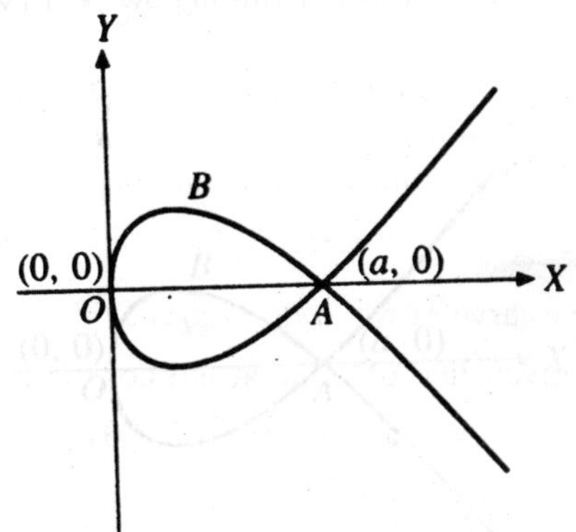

Fig. 3.49

Now $(dy/dx) = 0$ when $3x^2 - 4ax + a^2 = 0$ or when $x = a/3$ which gives from (1), $y = (a\sqrt{2})/(3\sqrt{3})$ *i.e.*, $< a$ showing that the loop does not intersect the straight line $y = a$.

By symmetry the C.G. of the loop lies on x-axis *i.e.*, the distance of the C.G. will describe a circle of radius a whose perimeter is $2\pi a$.

Also the area A of the loop

$$2\int_0^a y\,dx$$

$$= 22\int_0^a \frac{(x-a)\sqrt{x}}{\sqrt{(2a)}}dx,$$

$$\left[\because \text{from}(1)y = \frac{(x-a)\sqrt{x}}{\sqrt{(2a)}}\right]$$

$$= \sqrt{\left(\frac{2}{a}\right)}\int_0^a (x^{3/2} - ax^{1/2})dx$$

$$= \sqrt{\left(\frac{2}{a}\right)}\left[\frac{x^{5/2}}{5/2} - \frac{ax^{3/2}}{3/2}\right]_0^a$$

$$= \frac{4}{15}\sqrt{2}a^2.$$

$\therefore$ by Pappus theorem, the required volume

$$= 2\pi a \times A = 2\pi a \times \frac{4}{15}\sqrt{2}a^2 = \frac{8}{15}\sqrt{2}\pi a^3.$$

Example 81:

Find the volume of the ring generated by the revolution of the cardioid $r = a(1 + \cos\theta)$ about the line $r\cos\theta + a = 0$, given that the centroid of the cardioid is at a distance 5a/6 from the origin.

Solution:

The given curve is $r = a(1 + \cos\theta)$. ...(1)

And the given line of rotation is $r\cos\theta + a = 0$

or $x + a = 0, (x = r\cos\theta)$

or $x = - a$.

By symmetry the centre of gravity the centre of gravity G of the cardioid lies on the initial line OX. If G be the centroid of the area of the cardioid, then OG = 5a/6 (given).

Also GM = the length of the perpendicular from G on the line of rotation

$$= GO + OM = (5a/6) + a = 11a/6.$$

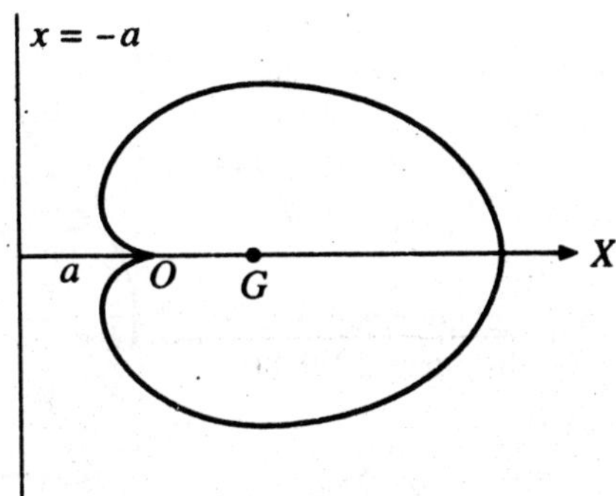

Fig. 3.50

Also the area A of the cardioid $= 2 \times \int_0^{\pi} \frac{1}{2} r^2 d\theta$

$$= \int_0^{\pi} a^2 (1 + \cos\theta)^2 \, d\theta, \text{ from (1)}$$

$$= a^2 \int_0^{\pi} \left(2\cos^2 \frac{1}{2}\theta\right)^2 d\theta$$

$$= 8a^2 \int_0^{\pi/2} \cos^4 \phi \, d\phi,$$

putting $\theta/2 = \phi$

so that $d\theta = 2d\phi$

$$= 8a^2 \cdot \frac{3}{4} \cdot \frac{1}{2} \cdot \frac{1}{2} \pi = \frac{3}{2} \pi a^2.$$

$\therefore$ by Pappus theorem, the required volume

$$= (2\pi . GM) \times A = 2\pi (11a/16) \times \frac{3}{2} \pi a^2 = \frac{11}{2} \pi^2 a^3.$$

Example 82:

Show that the volume of the solid generated by the revolution of the curve $(a - x)\, y^2 = a^2 x$, *about its asymptote is* $\frac{1}{2} \pi^2 a^3$.

Solution:

The given curve is $(a - x)\, y^2 = a^2 x$. ...(1)

Its shape is as shown in the figure.

Equating to zero, the coefficient of highest power of y, the asymptote parallel to the axis of y is $a - x = 0$ *i.e.*, $x = a$.

Take an elementary strip PMNQ, where P (x, y) and Q (x + δx, y + δy) are two neighbouring point on the curve and PM, QN are perpendiculars to the asymptote from the point P and Q respectively. We have PM = OL – OK = a – x and MN = δy.

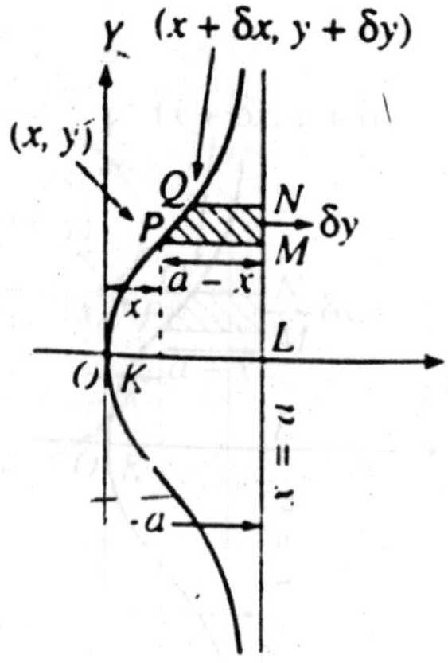

Fig. 3.51

Now volume of the elementary disc formed by revolving the strip PMNQ about the line x = a is

= p.PM^2.MN = π $(a - x)^2$ δy.

The given curve is symmetrical about x-axis and for the portion of the curve above x-axis y varies from 0 to ∞.

∴ the required volume

$$= 2\int_{y=0}^{\infty} \pi(a-x)^2\,dy$$

$$= 2\pi\int_0^{\infty}\left(a - \frac{ay^2}{y^2+a^2}\right)^2 dy, \qquad \left[\because \text{from}(1), x = \frac{ay^2}{y^2+a^2}\right]$$

$$= 2\pi\, a^6 \int_0^{\infty} \frac{dy}{(y^2+a^2)^2}.$$

Now put y = a tan θ

so that dy = a $\sec^2\theta\, d\theta$.

When y = 0, θ = 0

and when y → ∞, θ → π/2.

Therefore the required volume

$$= 2\pi a^6 \int_0^{\pi/2} \frac{a\sec^2\theta d\theta}{a^4\sec^4\theta}$$

$$= 2\pi a^3 \int_0^{\pi/2} \cos^2\theta d\theta$$

$$= 2\pi a^3 \cdot \frac{1}{2}\cdot\frac{1}{2}\pi = \frac{1}{2}\pi^2 a^3.$$

Example 83:

The figure bounded by a quadrant of a circle of radius a and tangents at its extremities revolves about one of the tangents. Prove that the volume of the solid generated is

$$\left(\frac{5}{3}-\frac{1}{2}\pi\right)\pi a^3$$

Solution:

Let the arc AB be the quadrant of the circle

$$x^2 + y^2 = a^2.$$

The area bounded by the arc AB and the tangents AC and BC (*i.e.*, the area ABCA) is revolved about the tangent AC, (say). Take an elementary strip PMNQ where P (x, y) and Q (x + δx, y + δy) are two neighbouring points on the arc AB and PM, QN are perpendiculars from P and Q on the tangent AC.

We have $\quad PM = OA - OK$

$$= a - x$$

and $\quad MN = AN - AM$

$$= y + \delta y - y = \delta y.$$

Now volume of the elementary disc formed by revolving the strip PMNQ about the tangent AC (*i.e.*, the line x = a) is

$$= \pi.PM^2.MN$$

$$= \pi\,(a - x)^2\,\delta y.$$

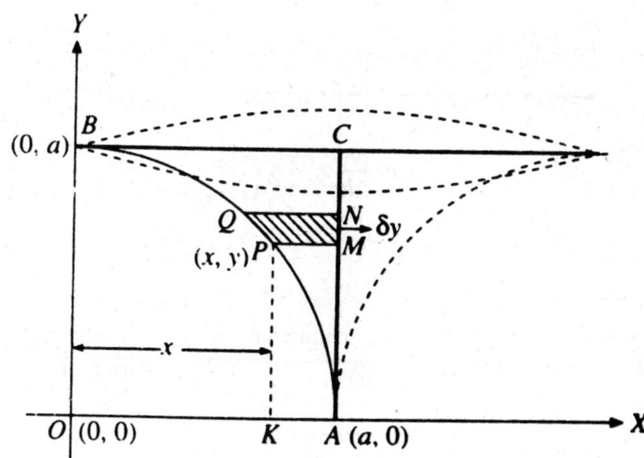

Fig. 3.52

Also for the arc AB, y varies from 0 to a.

$\therefore$ the required volume

$$= \int_{y=0}^{a} \pi(a-x)^2\,dy$$

$$= \pi\int_0^a (a^2 + x^2 - 2ax)dy$$

$$= \pi\int_0^a \{a^2 + (a^2 - y^2) - 2a\sqrt{(a^2 - y^2)}\}dy, \qquad [\because x^2 = a^2 - y^2]$$

$$= \left[2a^2y - \frac{y^3}{3} - 2a\left\{\frac{1}{2}y\sqrt{(a^2 - y^2)} + \frac{1}{2}a^2\sin^{-1}(y/a)\right\}\right]_0^a$$

$$= \pi\left[2a^3 - \frac{1}{3}a^3 - a^3\sin^{-1}(1)\right]$$

$$= \pi a^3\left[\frac{5}{3} - \frac{1}{2}\pi\right].$$